Applied Agricultural Biotechnology

Applied Agricultural Biotechnology

P Smith

2026

Agri Horti Press

New Delhi-110 084

ISBN 9789383285747 (HB)

Published By

Agri Horti Press

New Delhi

E-mail: agrihortipress@gmail.com

Digitally Printed at: Replika Press Pvt. Ltd.

Preface

Agricultural biotechnology companies work to supply farmers with tools to increase the yield of plant and animal products, while lowering the costs of production. Agricultural biotechnology can also include production of plants such as orchids for ornamental purposes and plants that can be used for fuel production .

Agricultural biotechnology is a collection of scientific techniques, including genetic engineering, that are used to create, improve, or modify plants, animals and microorganisms. Using conventional techniques, such as selective breeding, scientists have been working to improve plants and animals for human benefit for hundreds of years. Technological change in agriculture has been the foundation of economic growth and development. By raising yields and continually increasing the number of people who can be fed by the output of one worker in agriculture, change in agricultural technology has permitted an increasing proportion of labor and other productive resources to be devoted to non-agricultural production.

Every year, millions of people worldwide die from starvation and nutritional deficiencies. Although these challenges remain significant, we are making some progress. Agricultural technologies have increased caloric intake on a per-capita basis in nearly every nation of the world. Technological development—particularly in the area of agricultural biotechnology — is crucial if we want to further reduce malnutrition and starvation and meet the needs of growing populations. Unfortunately, misperceptions about biotechnology have led to activist movements to slow down, if not eliminate, this critical technology via government regulation.

Technologies, by definition, are neither acts of God nor nature; they are the embodiment of specific human purposes and intentionality. Far from being inevitable, they are researched and developed by those entities with sufficient power to mobilize social institutions to bring the technology into being, according to their own goals and normative considerations. 'Biotechnology is not neutral. This book sums up information about agricultural biotechnology and is a valuable tool for students as well as teachers.

Author

Preface

Contents

1

Introduction

Technological change in agriculture has been the foundation of economic growth and development. By raising yields and continually increasing the number of people who can be fed by the output of one worker in agriculture, change in agricultural technology has permitted an increasing proportion of labour and other productive resources to be devoted to non-agricultural production. Plant and animal breeding and chemical and mechanical technology plus improved husbandry have caused continual structural change in farming.

These technological stimuli have arisen as a series of overlapping waves to create a process of technological change, the momentum of which looks likely to be maintained by intensified adoption of information and biotechnologies. As a result of the cumulative processes to date, farms have become larger and more specialized, and farmers are more highly trained and have substituted machines for animal and human power; farming in the 'West' has become one of the most capital-intensive of industries.

At the same time agriculture has become more dependent upon industry, and even more industrialized-the intensive production of eggs, broiler chickens, horticultural products and pigs has many characteristics which belong to industrial production rather than to traditional farming. Goodman *et al.* talk of 'the industrial appropriation of the rural processes'. Processes of marketing farm products have passed off the farm to be performed by sophisticated distribution and retailing sectors which increasingly dictate details of production to the farm sector, while at the other end of the chain an increasing proportion of inputs is supplied by the chemical, pharmaceutical and engineering industries.

The process of 'industrializing' agriculture has been transferred to the less-developed countries (LDCs) and is epitomized by the so-called 'Green Revolution'. This has entailed expansion of irrigation systems using modern pumps, engineered dams and canals, plus the increased use of inorganic fertilizer, insecticides, herbicides, fungicides and power machinery; although at its heart has been the 'old' biotechnology of breeding new cereal varieties which give high yields when supplied with chemical inputs. Because much of the impetus for this 'revolution' has been western technology and farming know-how, particularly that of the USA, there are those who are critical of what they see as the LDCs' increasing technological dependence upon the West, its industries and research institutions.

What is undeniably the case is that technological advance has been faster, and public and private expenditure on research in western agriculture greater than in LDCs taken as a whole. Thus exportable surpluses from the West have increased with generous public support, while many LDCs have had to resort increasingly to food imports to meet their needs.

While at this stage it is difficult to foresee exactly when, and on what commercial scale, the new agricultural biotechnology 'revolution' which is now brewing will have its effect, most commentators agree that it will intensify the industrialization of agriculture, and that it will increase the technological dependence of most LDCs upon western firms. It may cause considerable disruption to the economies of some countries and will cause farming operations to diverge increasingly from the pattern associated with the country yeoman. Rural life in many more remote farming areas will be further threatened by the continued deterioration in the economics of ruminant livestock farming on upland and low-productivity pastures. It is these redistributive and structural effects that are the central focus here rather than any attempt to estimate the scale and speed of uptake of specific biotechnology inputs and processes.

Although prospective biotechnological change in agriculture is not different in many fundamental respects from other processes of technological change, in that it will operate through the provision of new or modified inputs (seeds, hormones, etc.) and the creation of new markets in industry, it does raise new issues. Are there dangers from releasing life forms which are engineered using new methods which are quite different from those associated with varieties developed by traditional plant-and animal-breeding methods? Even if the dangers can be objectively assessed and turn out to be minimal, will what may be perceived as man-made and therefore 'unnatural' products be acceptable to society as a whole?

Significant of Agricultural Biotechnology

Every year, millions of people worldwide die from starvation and nutritional deficiencies. Although these challenges remain significant, we are making some progress. Agricultural technologies have increased caloric intake on a per-capita basis in nearly every nation of the world. technological development— particularly in the area of agricultural biotechnology — is crucial if we want to further reduce malnutrition and starvation and meet the needs of growing populations. Unfortunately, misperceptions about biotechnology have led to activist movements to slow down, if not eliminate, this critical technology via government regulation. In the United States, the advancement of biotech crops has slowed down because of attempts to regulate them differently than conventionally grown foods. In Europe, antibiotech movements are much stronger, and international bureaucrats are pushing various international agreements and trade restrictions. While Western nations may be able to absorb the costs of such policies, people in the developing world will pay a higher price as many continue to suffer from starvation.

What is agricultural biotechnology?

Broadly speaking, biotechnology is any technique that uses living organisms or substances from these organisms to make or modify a product for a practical purpose. Biotechnology can be applied to all classes of organism-from viruses and bacteria to plants and animals-and it is becoming a major feature of modern medicine, agriculture and industry. Modern agricultural biotechnology includes a range of tools that scientists employ to understand and manipulate the genetic make-up of organisms for use in the production or processing of agricultural products.

Some applications of biotechnology, such as fermentation and brewing, have been used for millennia. Other applications are newer but also well established. For example, micro-organisms have been used for decades as living factories for the production of life-saving antibiotics including penicillin, from the fungus Penicillium, and streptomycin from the bacterium Streptomyces. Modern detergents rely on enzymes produced via biotechnology, hard cheese production largely relies on rennet produced by biotech yeast and human insulin for diabetics is now produced using biotechnology.

Biotechnology is being used to address problems in all areas of agricultural production and processing. This includes plant breeding to raise and stabilize yields; to improve resistance to pests, diseases and abiotic stresses such as drought and cold; and to enhance the nutritional content of foods. Biotechnology is being used to develop low-cost disease-free planting materials for crops such as cassava, banana and potato and is creating new tools for the diagnosis and treatment of plant and animal diseases and for the measurement and conservation of genetic resources. Biotechnology is being used to speed up breeding programmes for plants, livestock and fish and to extend the range of traits that can be addressed. Animal feeds and feeding practices are being changed by biotechnology to improve animal nutrition and to reduce environmental waste. Biotechnology is used in disease diagnostics and for the production of vaccines against animal diseases. Clearly, biotechnology is more than genetic engineering. Indeed, some of the least controversial aspects of agricultural biotechnology are potentially the most powerful and the most beneficial for the poor. Genomics, for example, is revolutionizing our understanding of the ways genes, cells, organisms and ecosystems function and is opening new horizons for marker-assisted breeding and genetic resource management. At the same time, genetic engineering is a very powerful tool whose role should be carefully evaluated. It is important to understand how biotechnology-particularly genetic engineering-complements and extends other approaches if sensible decisions are to be made about its use.

Understanding, characterizing and managing genetic resources

Farmers and pastoralists have manipulated the genetic make-up of plants and animals since agriculture began more than 10 000 years ago. Farmers managed the process of domestication over millennia, through many cycles of selection of the best adapted individuals. This exploitation of the natural variation in biological organisms has given us the crops, plantation trees, farm animals and farmed fish of today, which often differ radically from their early ancestors.

The aim of modern breeders is the same as that of early farmers-to produce superior crops or animals. Conventional breeding, relying on the application of classic genetic principles based on the phenotype or physical characteristics of the organism concerned, has been very successful in introducing desirable traits into crop cultivars or livestock breeds from domesticated or wild relatives or mutants. In a conventional cross, whereby each parent donates half the genetic make-up of the progeny, undesirable traits may be passed on along with the desirable ones, and these undesirable traits may then have to be eliminated through successive generations of breeding. With each generation, the progeny must be tested for its growth characteristics as well as its nutritional and processing traits. Many generations may be required before the desired combination of traits is found, and time lags may be very long, especially for perennial crops such as trees and some species of livestock. Such phenotype-based selection is thus a slow, demanding process and is expensive in terms of both time and money. Biotechnology can make the application of conventional breeding methods more efficient.

Genomics

The most significant breakthroughs in agricultural biotechnology are coming from research into the structure of genomes and the genetic mechanisms behind economically important traits. The rapidly progressing discipline of genomics is providing information on the identity, location, impact and function of genes affecting such traits-knowledge that will increasingly drive the application of biotechnology in all agricultural sectors. Genomics sets the foundation for post-genomics activities, including new disciplines such as proteomics and metabolomics to generate knowledge on gene and protein structure, as well as their functions and interactions. These disciplines seek to understand systematically the molecular biology of organisms for their practical use. A vast range of new and rapidly advancing technologies and equipment has also been developed to generate and process information about the structure and function of biological systems. The use and organization of this information is called bioinformatics. Advances in bioinformatics may allow the prediction of gene function from gene sequence data: from a listing of an organism's genes, it will become possible to build a theoretical framework of its biology. The comparison across organisms of physical and genetic maps and DNA sequences will significantly reduce the time needed to identify and select potentially useful genes.

Through the production of genetic maps that provide the precise location and sequences of genes, it is apparent that even distantly related genomes share common features. Comparative genomics assists in the understanding of many genomes based on the intensive study of just a few.

For instance, the rice genome sequence is useful for studying the genomes of other cereals with which it shares features according to its degree of relatedness, and the mouse and malaria genomes provide models for livestock and some of the diseases that affect them. There are now model species for most types of crops, livestock and diseases and knowledge of their genomes is accumulating rapidly.

BIOTECHNOLOGICAL PRINCIPAL of Agriculture

Agricultural technology can be defined as biotechnology to include 'any technique that uses living organisms or processes to make or modify products, to improve plants or animals or to develop micro-organisms for specific uses-it focuses upon recombinant DNA and cell fusion technologies'. Longworth concurs with this definition, with the significant addition of tissue-culture techniques, an aspect of biotechnology which has great economic, and therefore social and political, potential impact.

Despite debates about whether these or any other definitions are adequate, they do touch on the main aspects of biotechnology which deserve to be considered. The aspect of this new biotechnology which most captures the imagination and stirs the greatest controversy is gene splicing or recombinant DNA techniques, which inspire researchers to consider the possibilities of producing reproducible animals and plants markedly different from current existing species, and referred to as transgenic species. Already in higher animals and plants recombinant DNA has produced transgenic forms which are being commercially exploited. Their agricultural significance is, however, so far limited, and commercial application is concentrated in highly profitable pharmaceutical and horticultural markets. Genes have been introduced into several animal species which alter their protein synthesis to enable transgenic sheep to produce insulin in their milk, and rabbits to produce interferon. A completely different application has originated in Denmark for salmon, where it has proved possible to introduce germplasm which enables the salmon's physiology to handle heavy metals, which are normally toxic, so opening up new locations for farm fisheries. Much current research is directed to conferring disease immunity on animals, and holds out the prospect of widespread commercial application.

In plants one achievement has been the transfer of genetic resistance to antibiotics in the petunia, and another has been the introduction of storage-protein genes from French bean plants into tobacco plants. As yet commercial progress with recombinant-DNA technology in plants appears limited, but extensive opportunities beckon, particularly because plant research is less restricted by the ethical and animal-welfare concerns which apply to research on transgenic animals. The most important commercial developments based on gene splicing have so far occurred with genetically much simpler microorganisms, and it is with these that the greatest short- to mediumrun commercial potential lies.

Already genetically engineered micro-organisms are producing a variety of hormones, vaccines, enzymes and other proteins. Important examples are the production of insulin, vaccines for neo-natal diarrhoea in calves and piglets, and the bovine growth hormone BST identical to that produced in cows which can stimulate a 20 per cent increase in milk yield. BST is already a source of problems for legislators in the European Community, and has provoked strong reactions from the media and milk consumers. As far as larger animals, and cattle in particular, are concerned, it is developments in embryo transfer and in many processes for manipulating reproduction which hold out the prospect of continuing increases in yields, feed conversion efficiency and general economic efficiency.

Already, apparently over 1 per cent of dairy calves in the USA are from embryo transplants, despite the high costs still associated with this procedure. Widespread adoption of these sophisticated technologies would further distance livestock farming from its traditional rural simplicity, and from the natural mating of animals to produce offspring. It would place technical demands upon the operators which favour large-scale company farms capable of supporting a range of highly trained specialists. Longworth identifies new techniques of tissue and cell culture as having 'the potential for enormous advances in crop improvement in the next couple of decades'. Those techniques 'can both increase the genetic diversity and greatly increase selection efficiency', and they permit innumerable plants to be reproduced asexually from single cells or small pieces of tissue.

The particular technique known as 'callus culture' has been used for many years to clone highly valued horticultural plants such as orchids, and cloning of cuttings is widely practised for tree crops such as tea and palm oil as well as by millions of gardeners for garden plants. According to Longworth 'cell culture' of single cells has unexpectedly, and so far inexplicably, resulted in plants with different properties being regenerated from the same clump of parent tissue. Sugar cane, maize and potato plants regenerated in this way have been found which are resistant to important pathogens.

This application of cell culture with the capacity to generate vast numbers of seedlings rapidly has the potential to simplify greatly the hitherto labourious procedures of plant breeding and selection. However, in terms of current and immediate commercial importance, it is through micro-organisms that biotechnology has its greatest impact. Microbial fermentation processes have been of great commercial significance for centuries, for example in bread, wine and cheese production, and there is the prospect of considerable development. Two recent examples of new processes indicate the sorts of impact that such developments can have.

In the late 1960s genetically engineered bacteria were developed which were able to digest corn starch to produce high-fructose corn syrup (HFCS) and which left as a residue corn gluten, which is now an important protein feed for livestock. HFCS has made considerable inroads in the United States and in Japan. It has largely replaced sugar as a sweetener in Coca Cola as well as many other food and drink products. This has helped depress sugar prices to Third World growers. In the European Community steps have been taken to prevent HFCS and other new powerful sweeteners from being produced and from undermining the market for domestically grown beet sugar, which is supported by the Common Agricultural Policy.

A second important application of microbial fermentation has been the production of ethanol from sugar cane in Brazil and from maize in the USA. The commercial viability of these processes is critically dependent upon the price of oil as the main non-renewable source of fuel, and, to date, massive subsidies have been required to maintain the Brazilian and USA ethanol programmes. Longworth states, however, that a new biotechnology, Sucrotech, is being patented, which will not only reduce the cost of producing ethanol from sugar cane, but will simultaneously produce fructose at a cost which will be competitive with HFCS and may thereby reclaim part of the sweetener market for sugar cane.

That such possibilities are in prospect indicates how volatile the future might be; biotechnology has tilted the competitive balance from sugar cane to maize, causing economic pressure and even disruption to sugar-cane-dependent economies and may in future switch it back again. In the long run the capacity to ferment fuels microbially from renewable agricultural feedstocks points to an important long-term reorientation of agriculture if non-renewable oil becomes uncompetitively expensive as a fuel for cars and feedstock for certain chemicals. It suggests that eventually there will be an increased emphasis on agricultural production of industrial feedstocks at the same time as continually increasing food output will be required to feed the expanding world population. All this will require considerable increases in agricultural productivity, to which biotechnology will increasingly contribute, but it will at the same time impose great strains on the natural environment and the structure of agriculture.

USES OF BIOTECHNOLOGY

Biotechnology-the sum of the technologies deriving from molecular and cell biology-will have utility in plant breeding beyond the prebreeding stages. Genetic transformation—non-sexual insertion of alien genes and/or gene regulating systems—is envisaged as eventually being applied, quickly and easily, to finished cultivars, to endow them with a series of useful new traits.

However, experience is beginning to show that genetic background affects expression of genes and gene systems that are transferred *in vitro,* as is also the case when they are transferred via regular sexual mechanisms.

Long periods of trial and error are needed to bring transformed lines into desired levels of expression. Thus, for some years to come, genetic transformation perhaps will be used only as a prebreeding technique, one in which alien genes are put into breeding pools of elite germplasm, following which standard plant breeding methods are used to extract good new varieties containing the alien genes. Cell culture techniques—particularly protoplast fusion techniques—will have utility for making new combinations of nuclear and cytoplasmic genes.

Immediate utility is seen for manufacturing new male sterility systems to make F1 hybrid cultivars, or to develop new kinds of herbicide resistance. With time and experience it may be possible to increase primary productive potential as well, by selecting and fixing optimum mitochondrial and chloroplast genotypes, balanced with proper nuclear genotypes. Such success seems far in the future, however, since at present little is known about optimization of nuclear, mitochondrial, and chloroplastic combinations. DNA technology will help plant breeding in a more traditional way, when information is further developed on maps of restriction fragment length polymorphisms (RFLPs) and their linkages with useful traits in crop plants.

In vitro tests, using RFLP markers, will allow near perfect accuracy in selecting for such useful linked traits. RFLP technology also will be used for precisely characterizing cultivar lineages and relationships, for positive identification of miss-identified or mislabeled cultivars, for characterizing gene frequencies of populations, and even to test hypotheses of landrace evolution, and global movement of crop genotypes. Perhaps the greatest contribution of molecular biology to plant breeding will be a deeper and more precise understanding of gene action and inheritance, of cellular physiology and biochemistry, and perhaps of whole plant physiology. With such deepened understanding, breeding and selection via standard sexual methods will be planned and executed with greater vision and economy. Such knowledge and execution will be, in part, a consequence of efforts to make genetic transform-nation a practical reality. The fall-out of knowledge from genetic engineering efforts, wisely utilized by observant biologists, will be of great benefit to plant breeding, and to plant culture in general.

DEVELOPMENT OF NEW CROPS

In addition to broadening the genetic base of established crops, genetic enhancement-and plant breeding in the sense of final cultivar development-can be used in two other ways: to develop new crops from heretofore uncultivated species, and to change old crops into new crops. Breeding of the first type (creating new crops where none existed before) usually is interpreted also to include development of advanced cultivars from landraces.

Usually the advanced cultivars are made for industrial countries and the landraces come from developing countries. Soybeans and amaranth thus are spoken of as "new crops" in the United States, even though they have been cultivated for centuries as landraces in China, and in Central and South America, respectively. Challenges are great and are about the same, for changing either wild species or landraces into advanced cultivars, able to withstand monocropping and intensive culture at high production levels. Selection for productivity inevitably requires narrowing the germplasm base, in the course of selecting plant architecture and physiology able to withstand crowding and the stress of high yields, and amenable to machine planting, weeding and harvesting. Planting the narrowly selected genotypes in large areas gives excellent access to specially adapted disease and insect pests, as already noted, and so the need for tolerance or resistance always is next to be added to the list of breeding priorities for new crops. Thus, genetic enhancement first narrows and then later must broaden the genetic base, in the course of new crop development. And as soon as users of the new product get used to larger quantities of it, they will want to standardize and then to change the quality of the product-as, in oil percentage, grain storability and hardness, or flavour and texture. The rule is simple: the more successful the crop, economically, the greater the demands on (and complaints about) the plant breeders.

Breeding of the second type-changing old crops into new ones-has been considered as a broad-scale possibility only since concepts of genetic engineering were developed. Ability to produce human insulin in bacteria gave rise to thoughts of producing insulin, or other useful primary gene products, in crop plants. Plants were envisaged as now able to turn sunshine, air and water (and soil nutrients) into high value specialty chemicals, instead of commodity chemicals like corn starch or soybean oil. Most of these ideas still are only ideas, but workers are proceeding to put some of the least complicated ones into action. Modification, via genetic engineering, of the ratios of fatty acid components of rapeseed oil is contemplated. Standard breeding to modify chemical components has been used in Canada already, to make rapeseed into a new crop: "canola."

Canola oil, unlike standard rapeseed oil, is essentially free of erucic acid and glucosinolates and therefore is safe for human consumption when used as a cooking oil. Another possibility derives from recent success in cloning the R gene (first identified by Mendel) in peas which provides the opportunity to identify and transfer, *in vitro,* genes giving new starch branching properties and thus, potentially, useful new kinds of specialty starches in old starch-producing crops such as maize or wheat. A word of caution is needed for those looking to replace major portions of commodity crop production areas with specialty-product new versions of the old crops. High value specialty products, by definition, require only small amounts of product to saturate the market. They won't replace very much of the production area now used for production of commodity crops. On the other hand, they will make intensive use of crop breeding and crop management skills. They are "brain-power intensive" crops, and as such, must command higher preunit prices—premiums—or they won't be grown.

Photolysis and Carbon Fixation

Photosynthesis

Photosynthesis is the means that primary producers (mostly plants) can obtain energy via light energy. The energy gained from light can be used in various processes mentioned below for the creation of energy that the plant will need to survive and grow. Photosynthesis is a reduction process, where hydrogen is reduced by a coenzyme. This is in contrast to respiration where glucose is oxidised. The process is split into two distinct areas, *photolysis* (the photochemical stage) and the *Calvin Cycle*. The diagram below gives a summary of the reaction, where light energy is used to initiate the reaction in its presence:

$$CO_2 + H_2O \rightarrow \text{glucose} + \text{oxygen}$$

Photolysis

This part of photosynthesis occurs in the *granum* of *a chloroplast* where light is absorbed by *chlorophyll*; a type of photosynthetic pigment that converts the light to chemical energy. This reacts with water (H_2O) and splits the oxygen and hydrogen molecules apart. From this dissection of water, the oxygen is released as a by-product while the reduced hydrogen acceptor makes its way to the second stage of photosynthesis, the Calvin cycle. Overall, since the water is oxidised (hydrogen is removed) and energy is gained in photolysis which is required in the Calvin cycle

The Calvin Cycle

Also known as the carbon fixation stage, this part of the photosynthetic process occurs in the *stroma* of chloroplasts. The carbon made available from breathing in carbon dioxide enters this cycle, which is illustrated below: Just like the *Kreb's Cycle* in respiration, a substrate is manipulated into various carbon compounds to produce energy. In the case of photosynthesis, the following steps occur, which create glucose for respiration from the carbon dioxide introduced into the cycle;

- Carbon from CO_2 enters the cycle combining with Ribulose Biphosphate (RuBP)
- A compound formed is unstable and breaks down FROM its 6 carbon nature to a 3 carbon compound called glycerate phosphate (GP)
- Energy is used to break down GP into triose phosphate, while a hydrogen acceptor reduces the compound therefore requiring energy
- Triose Phosphate is the end product of this, a 3 carbon compound which can double up to form glucose, which can be used in respiration.
- The cycle is completed when the leftover GP molecules are met with a carbon acceptor and then turned into RuBP, which is to be joined with the carbon dioxide molecules to re-begin the process.

The energy that is used up in the Calvin cycle is the energy that is made available during photolysis. The glucose that is made via GP can be used in respiration or a building block in forming *starch* and *cellulose,* materials that are commonly in demand in plants.

Limiting Factors in Photosynthesis

Some factors affect the rate of photosynthesis in plants, as follows

- *Temperature* plays a role in affecting the rate of photosynthesis. Enzymes involved in the photosynthetic process are directly affected by the temperature of the organism and its environment
- Light Intensity is also a *limiting factor,* if there is no sunlight, then the photolysis of water cannot occur without the light energy required.
- Carbon Dioxide concentration also plays a factor, due to the supplies of carbon dioxide required in the Calvin cycle stage.

Overall, this is how a plant produces energy which supplies a rich source of glucose for respiration and the building blocks for more complex materials. While animals get their energy FROM food, plants get their energy FROM the sun.

CONTROL OF BIOTECHNOLOGY DEVELOPMENT

The role of the public sector to undertake research and development of agricultural technology as a public service to firms which might profit from translating that R & D into a commercial product or process, to farmers who might profit from adoption of the technology, and perhaps most importantly to consumers at home and abroad who benefited from the lower prices resulting from greater abundance. This was particularly true of the phase of change dominated by improvements in plant and animal breeding, the 'old' biotechnology. While the public sector had a role in basic research for chemical and mechanical technologies, the benefits of research expenditure in these areas were easier to capture by private companies investing in R & D, so that increasingly the public sector has taken a smaller role in these areas although maintaining a strong regulatory role in regard to agricultural chemicals in particular. Where it is impossible to prevent others from escaping payment for the research costs, either because the product is easily copied (seeds which can be regenerated by farmers or other firms) or because proposals for new methods can be readily implemented, private firms are understandably unwilling to invest.

Machines, insecticides, fungicides and other manufactured inputs do lend themselves more readily to private exploitation. Nevertheless the returns to R & D in these products do depend upon the degree of difficulty potential competitors would have in copying the product. In some cases there are inherent technical difficulties in copying the process, or it would be prohibitively expensive, but in other cases it is the ability to obtain patent protection which creates legal barriers to potential competitors' ability to become 'free-riders' and which protects incentives to private investment in R & D. Traditionally, however, it has been impossible for plant and animal breeders to obtain patent rights for their products, which is one reason why public-sector R & D has remained so important in this area.

For, as under the European Patent Convention (EPC) of 1973, it has been judged impossible for plant varieties to satisfy one of the key criteria to qualify for a patent, namely proof of 'an inventive step'. The application of standard breeding practices to generate new varieties by crossing existing plants or animal strains has not been deemed to be invention. Thus under the EPC one set of exclusions from patentability is 'Plant or animal varieties or essentially biological processes for microbiological processes or the products thereof.' In the absence of patent rights plant-breeding firms in particular have worked hard to obtain other means of protection and royalties for their products.

The history of the development of Plant Breeders' Rights (PER) is presented by Mooney, and is of particular interest because of the concerns he expresses about the consequences of allowing the basic genetic stock, which underpins agriculture and hence our whole society, from becoming private rather than public property. While Mooney's concerns are expressed in relation to 'old' biotechnology varieties of plants, they are of particular importance with respect to the products of the 'new' biotechnology. For bio-engineered products are capable of meeting the inventive-step criterion of patentability, and both micro-organisms and transgenic animals have now been patented in the USA.

Before examining developments in the seed industry it is worth touching upon the implications of changes in research policy which stress increasing reliance upon private R & D to develop and exploit the new biotechnology and other technologies. In the UK government has decided that it should withdraw from funding what it terms 'near-market research', that is, research beyond the basic phase and which is preparation for commercial exploitation. This policy is based in part upon the arguments that industry should be investing more heavily in R & D, and that research strategies at the near-market stage should be driven by assessments of likely commercial success which can best be made by the firms involved, and will therefore lead to greater efficiency in allocating research funds.

This has led to the closure of a number of government-financed agricultural research institutes, the scaling down of others and the sale of the National Seeds Organization by auction to Unilever; Unilever won against competition with BP and other major public companies. This deliberate attempt to switch an increasing proportion of agricultural R & D expenditure from the public to the private sector carries with it a number of risks.

In the first place there is a controversy about whether there is under-investment in R & D so that the returns to extra expenditure are high, or whether the converse is the case. If there is under-investment, then withdrawal of public support for applied research will exacerbate this, as the private sector will only undertake R & D expenditure on those products and processes from which exclusive benefits can be captured by the investor.

Thus in relation to biotechnology in the USA, where again public support for applied biotechnology research is relatively small, Stallman and Schmid argue that there will be emphasis on technologies which are applied in factory conditions where secrecy and control can be maintained. For technologies which cannot be confined to factories, such as seeds, Stallman and Schmid state that 'Firms are also considering mechanisms which "scramble" the genome of a plant in the second generation', in order to prevent farmers from reproducing seed with the enhanced, engineered characteristics.

Clearly such actions are designed to frustrate the maximum spread of benefits from the technology and to maximize private profit for companies investing in research.

Another facet of commercially orientated biotechnology research is that it will aim at the most important crops and livestock products, those produced by the largest farming units, and those which are most heavily subsidized. In the latter case this will worsen the budgetary problems of adjusting agricultural policy in OECD countries. The probable neglect of minor crops, difficult habitats and small farmers means that the public sector will have a defined role in agricultural technology research, but one in which it will be relegated to the second division and where it is unlikely to prove successful in terms of the commercial yardstick of rates of return which is increasingly emphasized by public-research policy.

Genetically Engineered Plants

At a symposium at the University of Washington Law School in October 1993, when discussing the 'Future of Intellectual Property Protection for Biotechnology in the US, EC, and Japan', Joseph Straus of the Max Planck Institute (Germany) warned the participants against 'ethics and other irrational considerations'. Indeed, except for a very narrow band of 'moral dilemma' situations which are the stock-in-trade of professional biomedical ethicists, the ethical aspects of genetic engineering have been routinely ignored by government policy-makers and corporate technology promoters. My thesis is that it has been the corporate promoters and their governmental handmaidens who have been 'irrational' in their systematic refusal to acknowledge the environmental and ethical considerations of genetic manipulation.

Technologies, by definition, are neither acts of God nor nature; they are the embodiment of specific human purposes and intentionality. Far from being inevitable, they are researched and developed by those entities with sufficient power to mobilize social institutions to bring the technology into being, according to their own goals and normative considerations. 'Biotechnology is not neutral. It shares the propensity of modern materialistic science to desacralise, dominate and manipulate life. It reduces all living things to a mechanism which it can manipulate according to engineering standards.' Genetic engineering is a technological process for undertaking activities which do not, and cannot, occur in nature. In this sense it is perfectly legitimate, therefore, to label genetic engineering as 'unnatural'. The European Community has incorporated this concept in its definition of genetically modified organisms, and it has been adopted by the UN Environment Programm (in the working of its Fourth Expert Panel set up under the Biodiversity Convention as part of the follow through to the United Nations Conference on the Environment and Development held in Rio de Janeiro in June 1992): 'organisms in which the genetic material has been altered in a way that does not occur naturally by mating and/or natural recombination'.

Thus, hybrids and other modified organisms obtained by traditional breeding techniques are excluded from the definition of genetically modified organisms. It is important to note that the process as well as the products are novel. The promoters of the new technology believe that a concern for the process itself is specious, downplaying its novelty whenever confronted by discussions of regulatory oversight. However, one of the bases of the decision of the US Supreme Court in the *Chakrabarti* case (which, 5 to 4, upheld the patentability of genetically engineered microbes, and which has been unquestionably accepted by foreign governments as disposative of the issue of patentability of genetically engineered life forms) was that the element of 'novelty' required by the patent law was to be found in the process by which the microbe had been created. One reason why we cannot ignore the powerful novel aspects of the processes of genetic manipulation is because we are not omniscient as to what will, in fact, happen when we alter genomes.

Genetic manipulation is not like a child's game of Legos or Tinkertoy, in which parts can be rearranged or linked up with only simple mechanical and relatively predictable consequences. In calculating any risk from a transgenic organism, one should consider four elements: the host organism, the foreign genes, the interaction between the foreign genes and the rest of the genome, and the environment in which the organisms will be used.... [in regard to the last two elements] the literature contains many examples of genetic manipulations where inserted genes did not respond in their new environments the way they did in their old ones or where alterations with one part of the genome caused surprising activity in other parts of the genome. There is an element of arrogance to scientific assertions which assure us that the process itself poses no risks, since scientists know so little about actual ecosystems.

For example, I have been told by agronomists that over 80 per cent of the organisms which can be identified in a soil sample from my garden are completely unknown to the scientific literature. Indeed, the Ecological Society of America itself has specifically warned about the problems presented in our lack of ecological knowledge and the resulting consequences from releasing (intentionally or accidentally) genetically altered organisms into ecological systems.

Foods

The Food and Drug Administration, responding to corporate pressures to remove the prospect of regulation of genetically altered foods, issued a set of rules which have largely left responsibility for protecting the public health and safety in the hands of the industry, permitting products to be marketed without scrutiny unless the industry indicated to the agency that it believed governmental oversight was justifiable. However, the proposal as published by the FDA in the Federal Register perversely noted several important problem areas: creating allergens, additions of genes from sources which might violate religious and cultural norms (of vegetarians, Jews or Muslims, etc.), and implications for animal welfare and well-being (for example, the incorporation of human growth hormone-producing gene into a pig's genome produced a highly arthritic animal).

Indeed, the commissioner of the FDA, David A. Kessler, and his colleagues noted the possibility that genetically engineered food might 'contain high levels of unexpected, acutely toxic substances'. In order to begin to address ethical issues presented by the genetic modification of foodstuffs, we need to have some ideas of the purposes for which these modifications are performed. (For example, if we wish to approach an analysis on utilitarian cost-benefit grounds.) Despite a great number of general statements, human nutrition and hunger do not appear to be the actual driving forces behind the development of genetically engineered foods. Other than claiming increased shelf life (which can be indirectly related to nutrition in the sense of reducing spoilage or the consumption of foods that have begun to spoil), there is hardly any indication at all that genetic engineering is being directed to create nutritious substances out of non-nutritious ones.

There are few real instances of genetic engineering reducing the cost of foodstuffs, or increasing the quantities of foods actually available to populations that are hungry because of the non-existence of consumables (as opposed to hungry because they lack money to buy sufficient foods, no matter how produced). Indeed, early genetic manipulations seemed perversely designed to minimize the achievement of such goals: the creation of 'herbicide tolerant plants' which do not reduce the applications of dangerous chemicals to agricultural fields (which might occur if genetic manipulations of food crops were designed to make them resistant to insects, fungi, or disease) but instead permit higher levels of chemical application, or the introduction of recombinant bovine growth hormone to produce more milk at a time when developed countries suffer from milk gluts and have instituted programmes to kill cows and physically dump milk (milk as a commodity is price inelastic; quantity increases do not lead to price decreases).

The goals of genetically engineering foods seem to be more closely tied to increasing the economic gain and power of the corporations involved in food production (for example, the development of herbicide-tolerant farm crops has been led by corporations which manufacture herbicides), making production easier for corporations, reducing or altering packaging and transportation costs for the corporations, etc.

Environment

Industrial societies around the world are faced by massive pollution legacies from the previous emphases on chemical and nuclear industrial activities. The inability of the US Department of Energy to find a suitable site for a longterm nuclear waste repository illustrates our very poor track record for dealing with the scientific, technical, economic, and socio-political aspects of prior technological 'revolutions'.

The environmental problems posed by genetically engineered organisms are likely to be substantially more intractable than those posed by these earlier instances of pollution because genetic wastes multiply, migrate, and mutate. A genetically engineered organism once free in the environment is impossible to recall. (Impacts which are irreversible must be considered with higher scrutiny than those which can be undone.) Examples of environmental problems which need to be assessed include the risks of transgenic crops themselves becoming weeds; the risk of gene flow to wild relatives which might become weeds or pests; the growth of antibiotic resistance in species (particularly animals), since resistance genes are used as markers for the genetic engineering sites; the problem of exotic or non-native species taking over ecological niches (such as gypsy moth, starling, kudzu, rabbits in Australia, etc.) leading to extraordinary economic losses as well as ecological ones.

The restriction (rather than increase) of biodiversity by selective advantageous breeding of transgenic organisms as opposed to natural ones, or the predatory results of expanding exotic species (such as the Dutch Elm micro-organism destroying the American Elm and severely restricting the biodiversity of certain areas in the Northeast United States); health issues (both of plant and animal species, as well as humans — worker health and safety as well as community security); and the occurrence of the completely unexpected, the inability to eliminate uncertainty

For example, the late 1993 floods in the Mississippi valley included the flooding of a field of genetically engineered corn and the dispersal of this plant material to unknown sites within the thousands of square miles of downstream flood plain. Prospective ecological assessment is not being performed, despite the fact that its need has long since been recognized. And economic priorities, particularly those accruing directly to the promoters, as well as the nebulous spur of 'competition' with foreign countries, oftentimes overwhelms any interest in doing environmental impact analysis.

Developments in the seed Industry

Many major food crops upon which we depend originated and were first cultivated in Third World countries rather than in the industrialized countries where they are most productively exploited today and in which they have been improved by traditional plant breeding. The potato originated in South America, wheat in Ethiopia and the Near East, important maize varieties in South America, and rice in Asia. Moreover the so-called Vavilov centres, which contain the greatest variety of living species and which have been the source of much important gene material, are mainly in Third World countries. The wealth of the industrial countries owes much to their past ability to exploit agriculturally crops and animals originating in other countries.

In the colonial era the success of British scientists in smuggling rubber plants from Brazil to Sri Lanka, Singapore and Malaysia and the comparable more complex route by which coffee was introduced to Latin America from Ethiopia are instances of colonial powers obtaining significant wealth by exploiting plant material from the Third World. In one way or another the process by which institutions and firms in developed countries have continued to collect new plant and animal varieties from less-developed countries has continued. But what is still a major source of friction is the procedures by which legislative protection is developed to give companies in the richer countries commercial rights to exclusive exploitation of varieties developed from plants freely collected from other countries, and to extract high rates of profit from the sale of those varieties. Most extreme is the situation whereby under the USA Plant Patent Act of 1930, which covered asexually propagated plants such as certain fruits, flowers, ornamental shrubs and trees, it has been possible for plants discovered and smuggled out of other countries to be given patent protection in the USA. The key to this is that the Act accepted the unfeasibility of requiring proof of an 'inventive-step' for the granting of a patent, and employed the criterion that as far as can be determined the plant variety is new.

Since some novel variety found in the rain forests of South America may satisfy this criterion, and the source of the plant can be concealed from the authorities, the theft and smuggling of plants can be said to have been encouraged. Seed companies in developed countries have successfully pressed their case for protection, and have progressively extended Plant Breeders' Rights (PBR) through both national legislation and the Union for the Protection of New Varieties of Plant (UPOV), which is an international agreement among signatory countries to honour a system of varietal rights.

In the case of food crops, in the absence of patentability, PBR operates by registering distinct varieties. Pressure has been exerted upon less-developed countries to become signatories of UPOV and thus acknowledge the rights of breeders, most of whom are from developed countries. However, the majority have stalled or rejected membership, and have been fundamentally unhappy about accepting the principle of creating monopoly rights for others over plant material which in some cases originated in their own country.

The whole issue of intellectual property rights has been raised to a prominent political level by the USA's insistence that it be included as one of the thirteen areas for negotiation in the Uruguay Round of multilateral bargaining of the General Agreement on Trade and Tariffs. While the issue is a much broader one than that of plant and animal breeders' rights, it reflects a basic political conflict between the richer countries, determined to obtain secure returns on R & D expenditures by companies, and LDCs struggling to catch up and anxious to avoid having to pay royalties on products protected by PBR and patents. This conflict has broader moral and ethical dimensions when it relates to the issue of charging royalties for the seeds of food crops to countries where malnutrition is widespread.

Against the previous argument has to be balanced the need of those incurring R&D expenditure to capture enough of the returns to provide an incentive for R & D. At least that is certainly the case where R & D is to be undertaken by the private sector, which is politically the increasingly preferred solution in the UK, the USA and elsewhere. This, however, leads to another area of concern, which is that a small number of large multinational chemical and pharmaceutical firms have increasingly come to control the seed industry. Companies such as Shell, Sandoz, Dekallb-Pfizer, Ciba-Geigy and BP have become owners of most of what were small independent seed companies.

This linking between chemicals and seeds within firms has a strong commercial logic. It facilitates the marketing of packages consisting of seed varieties plus the appropri-ate fertilizers, insecticides, etc. It also allows the companies concerned to integrate their plant breeding, biotechnology and chemical research. One particular synergy for these companies will be to bioengineer varieties of major food crops to resist their own herbicides and weedkillers. This would permit chemical weed control to be extended to large-scale crops where this is not now possible, and is just one way in which the direction of biotechnolo-gical research may be influenced (biased) by the combination of forces giving a small number of companies extensive control over the basic means of agricultural production. There are many other issues which could be touched on regarding corporate influence over the development of agricultural biotechnology.

It is, however, understandable that the spokesmen for LDC countries, which are profoundly concerned about their dependence upon western technology and exercised by what is seen as colonial and post-colonial exploitation by the West, should be alarmed by the prospect of new rights being created over seeds and other biotechnological products. Certainly there are widespread fears that they will only gain access to the fruits of biotechnology on unfavourable terms and that their relative dependence on and subservience to western industry will be increased so that benefits to them will be small. There are many western critics who would agree with this, and that the system by which agricultural biotechnology is delivered will favour the 'haves' rather than the 'have-nots'.

AGRICULTURAL BIOTECHNOLOGY REGULATION

For traditionally bred plants, regulators rely on plant breeders to conduct appropriate safety testing and to be the first line of defence against genetic alterations that might prove dangerous. The same should be true for agricultural products developed using biotechnology. There is no reason for regulators to treat biotech plants any differently than traditionally bred plants, particularly given the fact that biotechnology provides greater control over gene manipulation. Despite fears about gene manipulation, traditional crossbreeding has altered plant genes and improved the human diet for the past 30 years. In fact, people worldwide safely consume these plants every day. Examples include wheat, corn, potatoes, tomatoes, and countless other staples of the American diet. In 1984, the biotechnology industry began experimenting with "gene-spliced plants" — the more advanced approach to gene manipulation that we now call "biotechnology."

Department of Agriculture

USDA regulates the release of all genetically engineered agricultural plants under statutes giving USDA's Animal and Plant Health Inspection Service (APHIS) the authority to regulate plants that may be or may become weeds or other nuisances — what the statutes call "plant pests." Although the rules apply in a general sense to novel or exotic varieties of both gene-spliced and conventional plants, APHIS typically only requires field testing of conventional plants that are new to a particular U.S. ecosystem (transplanted from another continent, for example).

However, all genetically engineered agricultural plants face a higher regulatory bar — making it more difficult to expand the use of biotechnology. Genetically engineered crops must be field tested under APHIS regulations prior to commercialization. For example, a new variety of corn produced with conventional hybridization requires no government-mandated field testing, but all new varieties of genetically engineered corn do, even though there is no logical reason for the regulatory disparity. For most genetically engineered plants, APHIS requires the company producing the plants to submit notice detailing the gene or genes that have been inserted, where the plant will be tested, and other relevant characteristics of the plant prior to receiving permission to conduct the field trials. Once the company completes field testing, APHIS reviews the results and makes a determination on whether or not the product should be "deregulated" and can be released into the market.

Regulatory Scheme

At the time, the White House Office of Science and Technology Policy began crafting a framework wherein existing federal agencies would regulate genetically engineered organisms on the basis of their characteristics, not the method of production — thus wisely deferring to the scientific community's judgement that regulation ought to address the products, not the process of biotechnology. Engineered organisms would not require extra scrutiny simply because genetic engineering produced them (the process). Instead, they would be subject to heightened scrutiny only if the individual organisms expressed characteristics that posed some conceptually heightened risk (the product). The federal government divided regulatory jurisdiction among agencies already involved in agricultural, food, and environmental regulation. These include the United States Department of Agriculture (USDA), the Environmental Protection Agency (EPA), and the Food and Drug Administration (FDA). While each of these agencies considers the characteristics of individual products in their regulation, only FDA followed the general scientific thinking that genetically engineered and non-genetically engineered products should be regulated similarly. Both USDA and EPA automatically subject all genetically engineered plants as a class to premarket approval requirements not ordinarily applied to conventionally bred plants.

Environmental Protection Agency

EPA regulates plants that are modified to produce substances that act like pesticides: that is, substances used by a plant to protect itself from pests, such as insects, viruses, and fungi. EPA's proposed rule for these "plant-incorporated protectants" is not yet finalized. In the interim, however, EPA has regulated such plants by applying its proposed guidelines, which are functionally similar to rules for the registration of synthetic chemical pesticides. Again, biotech products face higher regulatory hurdles, even though plant-incorporated protectants developed through conventional breeding are exempt from these requirements.

Health and Human Services: Food and Drug Administration

FDA is responsible for ensuring that food items, including foods derived from genetically modified plants, are safe to eat. Following the general regulatory framework that emphasizes product regulation rather than process regulation, FDA rightly does not treat foods derived from genetically engineered plants to be inherently unsafe. Food producers are not required to seek premarket approval from FDA unless there is a substantive reason to believe that the novel trait (or traits) in the food pose a safety question. The initial determination of safety is left to the producer, but FDA has encouraged producers to consult with agency scientists prior to marketing a food produced with biotechnology to ensure that the appropriate determination is made. Recently, FDA published a proposed rule that would require producers to notify the agency at least 120 days before marketing, which may be a signal that FDA could abandon its more reasonable approach in the future. In addition, FDA does not require labelling of foods derived from biotechnology unless the genetic insertions so alter the food that the common name no longer describes it adequately.

Examples of this phenomenon would include such alterations of the food product that would raise a safety issue or change the product's nutritional composition or its storage or preparation characteristics.

International Trade

While U.S. consumers do not appear to be strongly opposed to biotech foods (they, in fact, seem rather indifferent), a strong anti-biotechnology movement has arisen in several European countries over the last few years. The European Union (EU) has established strong restrictions on the commercial planting of genetically engineered plants, and European food processors and retailers are reluctant to import harvested agricultural products derived from biotechnology — thus jeopardizing the marketability of U.S. commodity grain exports. And the EU is now negotiating for strong restrictions on agricultural biotech products in important international agreements governing food safety and environmental protection. Very strong restrictions were included in the Biosafety Protocol, finalized in January 2000.

Perhaps, more important, though, are the Codex Alimentarius Commission standards for food safety. Without convincing scientific evidence that genetically engineered crop plants pose a heightened environmental or human health risk, restrictions on agricultural biotech imports could be challenged under the General Agreement on Tariffs and Trade and the World Trade Organization. Thus the EU has resorted to justifying its decisions on the basis of a risk-management philosophy known as the "precautionary principle," and the EU advocates inclusion of the precautionary principle in international environmental and food safety agreements (such as the Codex Alimentarius), as well as within GATT itself. Its inclusion would give nations greater latitude in restricting imports of U.S. agricultural products.

No single definition exists for the precautionary principle, but its general meaning is that when an activity raises threats of harm to human health or the environment, regulatory measures should be taken to prevent or restrict the activity even if the risks have not been demonstrated scientifically.

Although the EU asserts that the precautionary principle is an unbiased risk-management philosophy, critics have noted that its lack of definition and evidentiary standards makes it all too easy to abuse for the purpose of masking trade protectionism, and that its very approach to risk management is inherently flawed and may, in fact, increase net risk.. Nevertheless, the precautionary principle has already been included in several international environmental treaties, including the Convention on Biological Diversity and the Biosafety Protocol.

Labelling Issues

Some activists, however, argue that the government should mandate the labelling of all genetically engineered foods. They assert that consumers have a "right to know" how their foods have been altered, and that a mandatory label would best allow consumers to choose between genetically engineered and conventional foods. Biotechnology advocates have argued against mandatory labelling because such requirements raise food costs— something that mostly harms lower-income Americans and people on fixed budgets. Perhaps more important, while biotech products are not substantially different from other products, special labels would likely make consumers think these products were more dangerous.

Hence, rather than serving educational or "right to know" purposes, such labels promise to simply confuse consumers. A government-mandated label on all genetically engineered foods also would raise important First Amendment free speech issues.

In 1996, the U.S. Court of Appeals, in *International Dairy Foods Association et al. v. Amestoy*, ruled unconstitutional a Vermont statute requiring the labelling of dairy products derived from cows treated with a genetically engineered growth hormone, noting that food labelling cannot be mandated simply because some people would like to have the information. "Absent ... some indication that this information bears on a reasonable concern for human health or safety or some other sufficiently substantial governmental concern, the manufacturers cannot be compelled to disclose it."

In other words, to be constitutional, labelling mandates must be based in science and confined to requiring disclosure of information that is relevant to health or nutrition. Furthermore, consumers need not rely on mandatory labelling of biotech foods to truly have a choice. Real-world examples show that market forces are fully capable of supplying information about process attributes (including kosher and organic production standards) that consumers truly demand. The same can be said about non-biotech foods, and the FDA recently published proposed guidelines to assist producers in voluntarily labelling both genetically engineered and non-genetically engineered foods. Additionally, the USDA's newly published rule for organic certification necessarily excludes biotech products from organic food production. Consequently, consumers wishing to purchase non-biotech foods need only look for certified organic products. Even as underdeveloped nations clamor for biotechnology applications, and as countries like China continue to experiment with and use agricultural biotechnology, opponents of agricultural biotechnology in the West, particularly Europe, attack it as an unnatural process that will destroy the world, not better it. They argue that biotechnology should be heavily regulated, if not banned. Already, however, genetically engineered plants are subject to strict regulatory oversight that is equal to or greater than that advocated by the vast majority of scientific specialists.

Additional regulation will slow down research and development of genetically engineered crops, keep beneficial products off the market, and raise the cost of products that do make it to consumers. Furthermore, the inclusion of similar restrictions — or inclusion of the precautionary principle— in international agreements will greatly impact the international trade of agricultural goods and delay their introduction into the marketplace. Each of these problems could prevent this technology's benefits from being introduced to industrialized nations and, more important, the developing world.

FUTURE OF AGRICULTURAL BIOTECHNOLOGY

There are divergent views about the rate of uptake of new agricultural biotechnologies and their impact. Kalter and Tauer and the US Office of Technology Assessment anticipate comparatively rapid take-up, while others (Buckwell and Moxey and Farrington do not foresee much impact until the next century, that is until ten to fifteen years have elapsed. The caution of the latter commentators is probably justified, for there are various hurdles to be jumped before biotechnologies can gain acceptance. One hurdle to be overcome is the economic one.

Self-evidently there must be some economic advantage to farmers or agro-industry from adopting the technology. Frequently what captures the scientific imagination turns out to be economically unviable in use, although it is only with use that efficiency can be improved and costs brought down. A more demanding test still for biotechnologies will be to obtain legislative approval and public acceptance. These are intimately connected. In the case of machinery developments, provided safety standards are observed there are no legal impediments to developing new and improved machines, since there are no obvious problems of hazard to the public and therefore public acceptance. Chemical insecticide, fungicides and herbicides do, however, pose much greater problems because of concerns with toxicity, and elabourate testing and licensing procedures have evolved after a relatively haphazard set of procedures in the 1950s and 1960s, when widespread use was made of DDT and organo-phosphorous compounds without adequate recognition of their toxicity.

In significant part, because of the problems which have been experienced with agro-chemicals, the licensing of biotechnologies will inevitably be based on testing at least as stringent as that which now exists for chemicals. Indeed the licensing procedures are likely to be more stringent because of public and scientific concerns. Public acceptance of biotechnology in the food chain will be made more difficult by confusion about differences between biotechnologies. In the early 1980s there was a crescendo of concern about the use of steroids to promote faster liveweight gain in calves and cattle.

The public outcry eventually led to the banning of such hormones for meat animals; but the legacy is a profound distrust of and hostility to new products such as synthetic Bovine Somatrophin (BST), which has the capacity to increase the milk yield of those dairy cows which are 'relatively' deficient in what is a naturally occurring hormone. Although initial scientific results are favourable, and some minor doubts remain, the major influence leading the European Commission to impose a moratorium on the use of BST until the end of 1990 at least is concern about public perception; the Commission has stated 'It would be a serious setback to producers and to the Community's milk policy were present [positive] trends in consumption to be reversed as a result of adverse consumer reaction. With newly bioengineered plants and animals scientific and public concerns are emerging which are of a different order from those associated with previous biotechnology. One relates to the possibility that crop failures may become more frequent if biotechnology leads to a reduction in genetic diversity in crops being grown; such a tendency has already been observed in relation to the Green Revolution technology for wheat. There are also concerns about upsetting the balance of nature in unforseen ways, akin to the unanticipated consequence of introducing rabbits into Australia and then trying to control these by introducing myxomatosis.

Thus questions arise such as what happens if herbicide resistance introduced into a commercial crop plant transfers itself by cross pollination into a closely related weed species, or indeed if the herbicide-resistant crop should colonize wild habitats. There are, of course, more lurid and absurd notions bandied about in the popular press which are similar in nature to the attacks made on Darwinism. The questions, both absurd and real, raised in relation to agricultural biotechnology will undoubtedly slow the rate at which it is adopted. Nevertheless instances of adoption are increasing: BST in the USA, a bioengineered baker's yeast in the UK, bioengineered sheep producing insulin, etc. As these become more widespread and increasingly affect lower-valued, bulk agricultural products, so the impacts of biotechnology on structural change will increase. The balance of economic power will switch increasingly to industry, to high-technology large farms and against smaller farmers in disadvantaged regions and countries. That, unfortunately, is a seemingly inevitable consequence of what we consider to be economic progress.

2

Social Resistance to Agricultural Biotechnology

investments in biotechnology research and development

Over the last two decades, the life sciences industry has made enormous investments in biotechnology research and development; thrown tremendous energy into getting its genetically engineered (GE) crops approved, patented, and commercialized; and lobbied U. S. farmers and food producers to use them. For their part, the U. S., British, and other governments that envision the biotechnology sector as the wave of the future, and as a means of augmenting their national competitiveness, have strongly supported the industry and its efforts to commercialize (and normalize) these new technologies.

They have devoted considerable sums of money to biotechnology research, taken significant strides to deregulate the industry, and sought to promote the spread of U. S.-style intellectual property rights in the World Trade Organization. The U. S. government in particular has also promoted the dissemination of agricultural biotechnology in developing countries through the U. S. Agency for International Development. With so much economic and political muscle propelling them, it is not surprising that GE crops hit the ground running when they came onto the scene in the mid-1990s. But what is surprising is that the rapid growth in GE crop deployment has been matched by an equally remarkable (and perhaps historically unprecedented) proliferation of citizens' voices challenging the biotechnology industry on economic, environmental, cultural, and moral grounds. Indeed, long before transgenic crops made their way to the market, individuals and groups concerned about the dissemination of these new technologies were already questioning their safety, utility, and necessity.

In this chapter, we explore the character and impact of the new social movement against genetic engineering in agriculture (hereafter, the "antibiotech movement"), as it is unfolding in the North, which is the heart of the agricultural biotechnology industry and the main arena in which these technologies have thus far been deployed. We analyse the strategies social activists have employed in efforts to limit the deployment of biotechnology in food and agricultural production, and their effectiveness in achieving this goal. This antibiotechnology activism, we contend, has turned what until recently looked like a done deal in the trajectory of agricultural industrialization—the shift from farming to biotechnology, as David Goodman and his colleagues have put it — into a moment of uncertainty and openness. Activism has created this uncertainty, we argue, by affecting the fortunes of the biotechnology industry and the political environment in which it is developing in three substantial ways.

First, activism has forced states to reconsider their liberal regulatory approaches to agricultural biotechnology, and to take steps to regulate the industry more seriously. Although pushing recalcitrant states towards stronger regulatory oversight remains difficult, there are signs that morestringent national-level regulation will be forthcoming in many countries, including the United States. Second, antibiotech activism is helping to spawn new global regulatory regimes for genetically engineered organisms, such as the Cartagena Protocol on Biosafety and new Codex Alimentarius Commission standards on bioengineered foods. Although social activists are not the main actors negotiating these regimes, they have helped ensure that their content moves the world towards more substantive regulation of these organisms. Third, the antibiotechnology movement has imposed important direct and indirect costs on the firms developing this technology. In addition, organized social activism has moved the issue of agricultural biotechnology out of relative obscurity, and out of the hands of a small number of corporate and state actors, into the public arena, where it is being debated by a broader spectrum of society.

In making this argument, we want to emphasize two points with respect to the recent sociological (as well as popular) literature on the industrialization of agriculture, of which biotechnology can be seen as the latest manifestation. First, in contrast to those who suggest that the events of the past several decades have provided transnational corporations with virtually unchallenged power to determine the future of world agriculture, we argue that social activists and their organizations are critical actors that have been overlooked by analysts of world-food-system restructuring. Without exaggerating their power with respect to large corporations and powerful states, we argue that social activists are playing an underappreciated role in shaping patterns of food production, consumption, and distribution. In the case considered here, antibiotechnology activists have placed new demands on biotechnology firms and, even more significantly, on states, demands that have forced them to abandon certain paths and move in new directions. Second, we want to distinguish ourselves from those who see consumer resistance as the primary basis for the recent change in the political and economic climate surrounding agricultural biotechnology. Although we do not mean to ignore the influence of consumer opposition, we hold that it is social-movement mobilization that has made consumers aware of genetic engineering of the food supply and that has motivated consumers to take a stance.

Even in Europe, where consumer opposition has been energized by food and health scares such as mad cow disease in Britain, AIDS-tainted blood in France, and a dioxin scare in Belgium, it is activist non-governmental organizations (NGOs) that have made agricultural genetic engineering into a social issue. We begin by providing a sketch of what the antibiotech movement looks like and who it comprises. We then discuss the three main arenas in which the movement has concentrated its efforts in the North—the domestic regulatory sphere, the international regulatory sphere, and the corporate sphere—and evaluate its effectiveness. Finally, we consider the strengths and limitations of this type of activism.

Biotechnology Innovation System in India

The government plays an important role in building the National Innovation System (NIS) by setting up institutions (public research institutions such as universities, labs etc), developing policies that support innovation and providing regulatory frameworks that can have a considerable influence on sectoral development. In the context of the developing countries, the role of the government becomes more prominent as the majority of the resources for the scientific research and development are being channelised or governed by the government. In case of India, nearly 80 per cent of the gross expenditure for research and development is being incurred by the Government while in case of the industrialized nations such as Japan, USA and UK the Industry spends nearly 75 per cent, 64 per cent and 44 per cent respectively. Though there has been a significant increase in the country's Gross National R&D expenditure in absolute terms from Rs.3974 Crore in 1990-91 to Rs.16199 Crore in 2000-01 to Rs.18000 Crore in 2002-03, the share of the government in national R&D expenditure has remained unchanged.

However, it is interesting to note that internationally all the governments perform the in-house R&D function, including even the highly private-sector-oriented governments such as the US and other OECD countries. A study of the Biotechnology innovation system of the four global leading countries namely USA, UK and Japan and Germany further supports the active involvement of Government R&D in furthering the growth of the Biotechnology sectoral innovation. Biotechnology is being characterised by multidisciplinary fields, high dependence on science and are emerging from industries having large R&D base, high degree of qualified manpower and having close links or interaction with public research institutes as well as universities and thus faces a broad set of different policy interventions by the government underlying the National Innovation System (NIS). The National System of Innovation (NSI) perspective is a convenient framework to understand the process of innovation occurring in an economy and especially within the manufacturing sector. It focuses on the country-specific factors influencing the process of innovation and economic growth. The NSI approach has evolved over the years both analytically and empirically with the contributions from Freeman, Lundvall and Nelson, Edquist (1997, 2002). NSI has became popular as an analytical framework to study innovation at the national, regional and even Sectoral level.

Though there are several definitions of NSI but that best fit in to the present context is that of Metcalfe (1995). NSI is defined by Metcalfe (1995) as a "set of distinct institutions which jointly and individually contribute to the development and diffusion of new technologies and which provides the framework within which governments form and implement policies to influence the innovation process. As such it is a system of interconnected institutions to create, store and transfer the knowledge, skills and artefacts which define new technologies".

NSI framework is comprised of various components such as the government, independent research institutes, firms and higher education system etc and their interrelationships. Innovation or government policy instruments and institutions play a very important role in cementing the relationship among the various components for the success of NSI. An application of this framework shall help in analysing the systemic failure that hamper the generation of innovations. Public policy can be applied to correct such failure.

There are several NSI studies on Biotechnology in the context of the developed world but the studies in the developing countries are rather limited. Biotechnology NSI in India has evolved over time and has been instrumental in creating infrastructure and capabilities in promoting biotechnology innovation in the country.

There are 241 institutions and 321 firms engaged in various dimension of biotechnology in the country. Indian Biotechnology industry is now among the top 12 countries in the world and like its global counterpart, is dominated by the health care sector. The recently released "Draft National Biotechnology Strategy 2005" by the Indian Government is a step forward in the direction of promotion and development of biotechnology to withstand global challenges and yet, solve the local health care needs of the country. Studies on biotechnology in India by and a recent empirical analysis of the health innovation system of seven developing countries further supports the role of the Governments in providing the sustainable environment for the successful development of health biotechnologies. According to Acharya *et. al.* government in developing countries is one of the key actors in prioritizing, implementing new technologies and promoting innovation to address their specific health and development needs. The governments of the countries where biotechnology innovation system has been successful in perform or support the following functions/ activities such as knowledge support, commercialization of research, collaborative research, financial support, regulation, IPR and resolving ethical issues etc.

Thus it would be interesting to analyse the role of the government in the promotion of the biotechnology in the country with special focus on the health care sector. Hence, the present study traces the role of government or national approaches in terms of policies, institutions and instruments in the promotion of biotechnology innovation system in India with a special focus on the health care biotechnology sector. A systemic framework similar to the EPHOITE (2003) project is being adapted to examine the direction of public policies, institutions and instruments in addressing the issues of knowledge support, commercialization of research, collaborative research, financial support, regulation, IPR and ethical issues in the area of health care biotechnology in Indian context.

The significant features of national biotechnology innovation system with focus on health care biotechnology with the help of various S&T indicators. The various government initiatives in terms of various policies, institutions and instruments in the framework of NSI while the lessons learnt along with the policy implications and challenges for the growth of health care biotechnology innovation.

Modern Agricultural Biotechnology

In 1973, Cohen and Boyer transferred a gene from one organism into another. In 1982, the first biotech plant, an antibiotic resistant tobacco, was developed. In January 1983, at a meeting of genetic researchers in Miami, three different teams reported success in using Agrobacterium tumefaciens, a bacterium, to carry new genes into plant cells, heralding the dawn of modern agricultural biotechnology. Agrobacterium tumefaciens, described as a "natural genetic engineer", splices its own genes into host plant cells. This pathogenic bacterium was now converted into a pack mule, to carry new, foreign genes into plant cells, minus the disease and this became the most common means of producing Genetically Engineered Organisms (GEOs).

Field tests for GE crops resistant to pests and pathogens were first conducted in the US, in 1985. A co-coordinated framework for the regulation of GEOs was established and the first GE tobacco was released, in 1986 . The US Department of Agriculture published guidelines for field trials of GE crops in 1991. On approval from the Food and Drug Administration, Flavour Saver, the first GE tomato, with a longer shelf life, was on the US markets in 1994. During 1995-96, GE soybean, corn and cotton were approved for commercialization, in the US. A number of GE crops, developed for pest, pathogen and herbicide resistance, are now commercially cultivated in several countries. Rice with pro-vitamin A, higher iron content, or human milk proteins and potatoes with high protein content are in various stages of development. Tobacco plants producing functional human haemoglobin and bacteria that produce human insulin have been developed, as well as plants with vaccines against rabies and other viral diseases. Food grain crops that withstand drought and salinity are high priority research, and so are those for high yield. A GE tobacco plant detoxifies soils contaminated by explosive residues, providing solution to a frustrating environmental problem in countries ravaged by armed conflict.

Now more than 70 biotech agricultural crops that have been approved for use in North America, including varieties of soybeans, cotton, canola, corn, potatoes, squash, tomatoes and papaya. About six million farmers in some 17 countries now cultivate GE crop on about 125 million acres, a 30 fold increase over 1996. By end of the year 2002, six GE crops planted in the US (soybeans, corn, cotton, papaya, squash and canola) produced an additional four billion pounds of food and fibre on the same acreage, improved farm income by US $ 1.5 billion and reduced pesticide use by 46 million pounds. In 2003 in the US, 80 per cent of soybean acres will be planted with biotech varieties.

A number of activist groups vehemently attack GE technology and its products on grounds of safety to humans and the environment, and costs of technology transfer and its reach to the needy. Products of agricultural biotechnology bear the brunt of this ill-informed, unscientific and prejudiced onslaught much more than GE products related to health or industry. In 2001, the European Community released results of a 15-year study, costing US $ 64 million, and involving more than 400 research teams and 81 projects. This report concluded that GE products pose no more risk to human health or the environment than conventional crops. So far, extensive and intensive research on the probable risks of GE technology has not brought out any adverse effects and none of the fears expressed by anti-tech activists were proved even marginally. Nevertheless, caution and examining issues case by case, is the watch word of technologists, who are aware of their responsibilities. For a number of products, technology transfer is free of costs for developing countries, as for example Golden Rice, the rice with pro-vitamin A. It is the responsibility of the Governments of the respective countries to bear the costs of developing local varieties and to reach the products to the needy at an affordable cost. In India, three GE varieties of pest resistant cotton were approved for commercialization, a year ago. Approval for other GE varieties of cotton and GE mustard was deferred twice by the Genetic Engineering Approval Committee (GEAC), the highest authority on the issue.

In India, the level of public awareness of the realisable benefits and probable risks of GE products is abysmally low. The functioning of the GEAC leaves much to be desired. Taking advantage of this hazy situation, mixing up economic, social and political issues with science, and even using such vagaries of nature as the severe drought, several groups of vested interest have created mistrust, confusion and scare. Trashing a very promising technology this way does not augur well for the future of Indian agriculture. This results only in denying the benefits of technology to the farmers and consumers.

FLUTTERING BENEATH THE MONARCH SUITS

In the mainstream U. S. media, one is most likely to encounter three rather different representations of antibiotechnology activism. One projects an image of playful public protest, such as the piece of street theater outside the Food and Drug Administration's (FDA) 1999 public GE hearing in Chicago, which featured a human fish-tomato, a grotesque cow that was shot up with genetically engineered bovine growth hormone by a mad scientist, and eight children dressed as monarch butterflies who were symbolically killed by genetically engineered mutant corn pollen. This sort of action has created a distinctive iconography for the antibiotechnology movement, in which the monarch butterfly has become a prominent symbol. Thus, in the popular street politics that has become commonplace in Northern countries since the 1999 WTO protest in Seattle, the antibiotechnology movement has a discrete recognizable presence that is not reducible to the antiglobalization sensibility into which many observers lump this politics.

In fact, many movement participants are associated with local foodsafety organizations or are organized in local or regional activist networks, such as the U. S. RAGE (Resistance against Genetic Engineering) and British GenetiX Snowball networks. They mobilize for specific campaigns and have played a major role in grassroots civic and protest activities such as "raiding" supermarkets and labeling GE products. The second representation is less forgiving. It depicts "special interest terrorists" and "eco-warriors" who clandestinely try to derail the biotechnology industry by vandalizing research facilities and destroying seed trials. Such actions are generally undertaken by local grassroots groups (some with rudimentary organizational bases) that often sport rebellious names, such as the Dusty Desperadoes and Strawberry Liberation Front in northern California, and GrainRage and the "elves" of the Earth Liberation Front in Minnesota.

By some accounts these groups represent no more than an alternative youth culture with anarchist leanings. They have fueled an image of nocturnal sabotage, illegality, and antisocial mayhem that has led legislatures in several U. S. states to press for tougher law enforcement penalties in property-destruction cases. Indeed, in February 2002, the Federal Bureau of Investigation singled out the Earth Liberation Front as a particularly dangerous domestic organization, one that poses a serious terrorist threat to the United States. They are, in fact, a fringe element in the movement, though they have been expensive to the industry and, in some people's opinions, to the movement as well.

The third image is one of antiscience, antiprogress Luddites, an image that to some observers also connotes a mysticist sensibility. This image is frequently invoked, especially by scientists, who express anger at the critics of biotechnology. A good example is Norman Borlaug's characterization of the "antiscience zealotry" that drives antibiotech activists. To many scientists, the extraordinary risk-aversion of biotech's critics and their insistence that we should step back from the potentials of scientific discovery simply because they fear there is a genie in the bottle are irrational, non or antimodern, and selfish. Yet those who frame the issue in this way fail to recognize that a significant segment of the antibiotech movement also comprises intellectuals, researchers, social scientists, and natural scientists working in established institutional settings on a range of issues.

Many antibiotech activists are scientific experts in their own right, such as Rebecca Goldburg at Environmental Defence, Jane Rissler and Margaret Mellon at the Union of Concerned Scientists, and Michael Hansen from Consumer Policy Institute. The movement also maintains connections with university scientists. The importance of this small but crucial group of activist-experts is that they are able to engage in the politics of expertise, framing their claims to authoritative knowledge in the terms used by industry scientists but presenting the issues from the perspective of the public interest. All these images of antibiotech activism, taken separately or together, are at best partial, and even misleading. What the images do indicate is that the movement is, both strategically and organizationally, diverse and complex. Strategically, the movement works in a rather loosely connected bricolage pattern of action: there are campaigns, labeling drives, protests, trial-crop destructions, and also serious research and knowledge-generation efforts to counter science with science. Organizationally, the movement comprises an extensive web of activist networks that engage a wide array of civil-society organizations connected not only across the world but also across a variety of issues, social concerns, and interest groups. This network mobilizes the resources of both mainstream and radical environmental NGOs, running the gamut from the Sierra Club and the British Soil Association to Greenpeace and Friends of the Earth International. It includes science based groups, such as the Council for Responsible Genetics in the United States; the Research Foundation for Science, Technology and Ecology in India; and the Institute for Science in Society in the United Kingdom.

It links the lobbying capacity of food policy NGOs such as the Institute for Development Policy ("Food First") and the Institute for Agriculture and Trade Policy in the United States with Southern activist groups such as the Tamil Nadu Women's Forum in India, Bio Watch South Africa, and MASIPAG in the Philippines. It engages farmers' organizations in the North as well as the South, such as the Confederation Paysanne in France and the New Agriculture Movement in Bangladesh. In short, the network nodes of antibiotechnology mobilization extend across local, national, regional, and transnational organizational structures. In a movement of this kind, organizational leadership and unity give way to coordination and decentralization. This means that local-level activism has cultural and political dynamics that go beyond street theater, mass protest, and the illegal destruction of trial plots. In Britain, the United States, and Canada, activists have been arrested inside supermarkets for affixing warning labels to products containing GE ingredients. In Ecuador and Brazil, activist groups have prevented ships carrying GE grain from docking. In the United States, local lobbying campaigns have persuaded city councils in Austin, Boston, Boulder, Ann Arbor, and other places to require GE food labeling.

Moreover, local actions are hooked into a larger panoply of activities facilitated by the network structure. In particular, activist organizations have used communication networks and the Internet to advertise protests and other actions, to generate petitions, and to solicit letters of protest against governments and firms around the world. Internet sites such as Resistance is Fertile and GenetiX Alert were established exclusively as clearinghouses of news and information about antibiotechnology actions.

Such informational networks have facilitated creative collaborations between NGOs and local activists across the North and South. The United States-based Institute for Agriculture and Trade Policy, for instance, supported a long march for sustainable agriculture in Thailand; and Greenpeace has been a ubiquitous campaigner, assisting local actions from

Boston to Bombay. In addition, these networks have enhanced the ability of NGOs and advocacy groups in the North to coalesce and share resources. In July 2000, for instance, seven major public-interest NGOs in the United States formed the Genetically Engineered Food Alert to press for a moratorium on genetically engineered foods until they are safety-tested and labeled and until biotechnology companies accept liability for any harm to human health or the environment.

The coalition's strength is its ability to coordinate the resources and skills of the different organizations while moving information rapidly between them in order to build public support and sustain pressure on the biotechnology industry. As with any social movement, the prospects of the antibiotech movement depend on its ability to maintain its momentum. Although the future is hazy, activists have become a real force, and this complex, multifaceted movement has real institutional strengths.

These strengths are manifested particularly in the ability of organizations to coalesce and to apply the weight of their resources and capacities—finance, human power, professional skills and training, infrastructure—synergistically to the pressure points of regulation and profitability in the deployment of agricultural genetically engineered organisms (GEOs). In so doing, these organizations engage two core principles of a democratic polity: accountability and choice.

limit the production and trade

In trying to limit the production and trade of GEOs, antibiotechnology activists have focused their efforts in three broad arenas. One is the domestic regulatory arena, where activists are pressuring the state to become a more serious regulator of biotechnology; the second is the international arena, where activists are working to create new supranational regulatory regimes for the global deployment of GEOs; and the third is the economic arena, where activists have sought to pressure biotechnology firms and consumers of their products to change their behaviour. In each of these arenas, antibiotechnology activists have developed carefully thought-out strategies and achieved significant successes.

Holding the State Accountable

Although some of the globalization literature claims that states have been rendered powerless by globalization, this is clearly not true with respect to the regulation of biotechnology. States continue to exert significant leverage over the biotechnology industry and its behaviour at the national level, particularly in the advanced industrialized countries where biotechnology has been most aggressively developed and deployed.

It is true that governments in some advanced capitalist countries—especially the United States, Canada, and Britain—have chosen to regulate the agricultural biotechnology industry only minimally and to encourage its growth. In this context, antibiotech activism has concentrated on exposing the dangers inherent in such a liberal regulatory approach and on pushing governments to adopt more comprehensive and responsible regulatory policies.

On both sides of the Atlantic, activists have revealed the weaknesses of existing regulatory structures by making embarrassing public revelations of government practices. In May 2000, for instance, activists in the United Kingdom seized on the fact that a Canadian company had inadvertently (and illegally) sold conventional rapeseed (canola) contaminated with Monsanto's Roundup Ready rapeseed to British farmers, who unknowingly planted it on more than ten thousand acres. Activists used the incident to claim that the technology is effectively uncontrollable and that the government was overly complacent in providing no oversight. (The problem had, in fact, been discovered not in Britain but in German field tests.)

In an effort to minimize the seriousness of this event, government politicians initially insisted that the contaminated seed was effectively sterile and unlikely to cross-pollinate. But activists' ability to expose the government's low level of knowledge, its weak regulation, and its tendency to take industry claims at face value severely undermined its popular credibility. Just four months later, antibiotech organizations in the United States made similar hay of the StarLink taco debacle, described in the introduction to this volume. In this case, activists pioneered the use of a new tactic: paying for independent testing of food products to see whether they contain genetically engineered ingredients. Indeed, it was the antibiotechnology coalition Genetic Engineering Food Alert that hired an Iowa-based genetics-testing firm, headed by a sympathetic scientist, to test whether several commercial brands of tacos shells contained StarLink.

After determining that the taco shells did indeed contain the unapproved corn, the coalition shared this information with the press and took the U. S. government to task for creating a regulatory system that could allow a bioengineered product to be approved for one type of use (animal consumption) but not another (human consumption), with no effective means of keeping the two separate. In addition to the tremendous economic costs that the StarLink debacle imposed on the industry, retailers, farmers, and grain traders, the incident pushed the issue of agricultural biotechnology—and the weakness of government oversight—squarely into the public eye. In both the rapeseed and taco cases, antibiotechnology activists stressed the difficulty of controlling this technology (especially under a free-market system) and highlighted the responsibility of the state to do so. They have complemented this strategy with sustained litigation. Court action is a powerful instrument for raising the stakes of public accountability in the foodregulation regime because it enables activists to frame specific and sometimes technical questions in terms of the state's responsibility to the public. For instance, in 1999, a coalition of activist organizations and citizens filed suit against the FDA, seeking to overturn its 1992 policy of neither labeling GE foods nor subjecting them to special testing. The litigants claimed that the FDA's decision violated the Federal Food, Drug, and Cosmetic Act, the agency's principal regulatory instrument. The point of the lawsuit was to show that the FDA had failed to ensure the safety of consumers regarding GE foods because the available scientific knowledge did not sustain the assumption of "substantial equivalence" and because safety tests carried out by the industry were tailored to meet government criteria that were designed to favour the industry.

Thus, the suit was not about the safety of GE foods *per se*, but about the adequacy and accountability of government regulatory procedures. One consequence of the suit was that it forced the government to place in the public domain a vast amount of information about internal disagreements, decision-making procedures, and the bases of scientific findings, information that compromised the confident tone of its regulatory pronouncements. Activists have also pressed for public accountability with respect to the environmental impact of the new agricultural biotechnologies. In 1997, seventy non-profit organizations filed a petition against the U. S. Environmental Protection Agency (EPA) in an effort to stop the release of crops modified with the bacterium *Bacillus thuringiensis*, or Bt crops. Supporters of the technology argue that Bt is a boon to the environment because its built-in resistance to pests enables farmers to use much less insecticide. Opponents argue that there is a good chance that pests will become tolerant to the Bt toxin, thereby reducing its effectiveness and its utility in organic farming. An additional concern is the potentially harmful effects on other species, such as the monarch butterfly larvae, that were indicated by a study at Cornell university. The petition called on the EPA to withdraw all current registrations and deny future approvals of Bt crops, pending appropriate evaluation of environmental risks and economic impacts on organic growers.

This petition was unsuccessful. But it demonstrates how such actions can increase public pressure on government agencies to exercise stronger oversight. As part of the petition, Greenpeace commissioned a review of the impact of transgenic Bt plants on non-target organisms. The review, carried out by an independent group of Swiss scientists, was highly critical of the industry's scientific tests, which provided the basis for the EPA's approval for insect-resistant crops.

Though the EPA continues to insist that its regulatory process ensures that these products are safe for human health and the environment (partly through a controversial and costly "refuge" requirement), the agency was forced onto the defensive and agreed to undertake a comprehensive review of its policies in 2000–2001. The FDA and the U. S. Department of Agriculture (USDA) also reviewed their agencies' regulatory procedures during this same period, largely in response to activist scrutiny and pressure.

In tandem with these "risk and procedure" cases, activists have also pushed "risk and rights" cases against state regulatory agencies. In particular, activists have argued that clear labeling of GE and non-GE foods would allow consumers to make informed choices about their own level of acceptable risk. Activists have pushed hard on this front because the political logic is inexorable: that citizens should have the right to know what they are ingesting is so obvious that almost any argument for curtailment advanced by the industry or by the government seems suspicious.

Indeed, repeated polls have indicated a widespread public preference for labeling. In March 2000, the Centre for Food Safety (CFS) spearheaded a large coalition of organizations that filed a legal petition with the FDA demanding the development of a thorough premarket environmental-testing regime and mandatory labeling for GE foods. The CFS built its argument on consumers' right to know which foods have been genetically engineered and their right to make an informed choice.

This tactic aimed to inject citizens' rights into the struggle over regulation as part of a broader strategy to force government agencies to articulate their responsibilities to citizens in the authoritative arena of the court and to pressure retailers to treat their consumers with respect. How effective have such actions been? In Europe, activist-inspired public pressure virtually forced the fifteen governments of the European Union to stop approving new GE crops in 1998—a de facto moratorium that effectively persists today. And in the United States, some fifty thousand public comments to the FDA (many of them the result of organized petition drives) and three heated public meetings in 1999 pushed that agency to develop a more stringent policy proposal for approving genetically engineered plants.

In 2000, then secretary of agriculture Dan Glickman initiated a review of the USDA's GE-related policies and procedures by the National Research Council of the National Academy of Sciences. The review was critical of current regulatory procedures, specifically calling for greater public participation and interdepartmental scientific review in the regulatory process. Though it is still too early to discern the actual impact of the independent review, it might lead to some policy changes.

Given the power and authority of the organizations involved in the development of biotechnology, these shifts are important markers of political struggles over the deployment of GEOs, especially in the United States, where industry strategists have relied on high levels of public inertia. To be sure, part of the motivation for reconsidering regulation has been the rising outcry against GEOs in Europe, the threats to trade, and the falling sales faced by biotechnology companies.

But these developments have been greatly advanced by the actions of public advocacy and environmental activist groups. In effect, activist initiatives have forced government regulatory agencies to respond to their efforts to shape a critical public consciousness of biotechnology's social impact and their efforts to force the state into a tighter regulatory role.

Social Activism in Putting the Industry on the Defensive

Social activism has not only forced the issue of biotechnology firmly on to the regulatory agenda but also complicated the economic lives of the biotechnology corporations. In recent years, church groups, consumers, farmers and farmers' associations, chefs, food processors, food retailers, and professional environmental and food-policy activists have all put direct pressure on these corporations, forcing them to rethink their strategies and, in some cases, to change their behaviour.

Perhaps most striking about the experience of the biotechnology industry in the 1990s is how unprepared it was for the strong reaction to its new attempts to "improve" nature. As one economist dryly observed, "Monsanto didn't even know a train was *coming* before it got run over." This may explain in part why the industry was so slow to react to the barrage of criticisms, consumer campaigns, lawsuits, and other actions that have thrown this techno-scientific project into question. But another part of the explanation surely resides in the divergent character—and preparation—of the two adversaries in this intense struggle for public opinion. On one side is a group of scrappy and diverse activist organizations who can punch and jab faster than you can say "genetic engineering," and on the other side is an elephant tethered to a post by its size, its enormous and sunk costs, and its competitive character. Moreover, activist organizations have spent the last twenty years in basic training for just such a battle, one that requires particular political, organizing, and issue-framing skills and occurs on many different fields simultaneously. By contrast, many life sciences corporations were so preoccupied with building up their strategic positions in this long lead-time, capital-intensive, and risky industry and with trying to control the regulatory agenda that they barely knew they had an adversary in the making.

From this position of strategic if not economic advantage, antibiotechnology activists and other concerned groups have launched an all-out offensive against firms situated at virtually every point along the commodity chain. They have instigated shareholder actions against publicly owned corporations, demanding that they disavow the use of genetically engineered inputs. Activists have also exerted pressure through lawsuits, such as one filed in 1999 by several prominent antitrust lawyers against Monsanto on behalf of small farmers and farm groups. The suit, inspired by a coalition of environmental groups, accused the biotechnology giant of being at the centre of an international conspiracy to control a large part of the world's seed supply and of giving farmers "false and fraudulent guarantees about the safety and marketability of a new breed of bioengineered seeds".

Activists have also organized consumer campaigns to convince food processors, retailers, and restaurant chains to stop selling genetically engineered foods, under the assumption that the market signals would quickly travel back up the supply chain to exporters, farmers, and, ultimately, the life sciences companies. In response to consumer activism, a number of major supermarket chains, particularly in Europe, started to remove GE products from their shelves as early as 1998, and some North American and Canadian chains are now following suit.

The Organic Consumers Association and the BioDemocracy Campaign have also targeted fifteen food companies and distributors—dubbed the "Frankenfoods Fifteen"— to get them to stop using GE crops in their products. In general, these campaigns have been surprisingly successful. Since 1998, some of the world's major food processors, including H. J. Heinz, Gerber, Bestfoods, Frito-Lay, Unilever, Seagram's, and Nestle, have publicly declared their products to be GE-free in some or all their markets. And in August 2000, a major biotechnology producer, Novartis, declared that it was eliminating GE ingredients from all of its consumer food products—in effect refusing to buy its own products. Consumer pressure—or the fear of it—has affected the whole food system. In 1999, McDonald's, fearing a threatened boycott, decided not to use genetically engineered potatoes in its French fries. After similar decisions were taken by Wendy's and Burger King, three major potato processors— J. R. Simplot, McCain Foods, and Lamb-Weston—advised farmers they would not be buying Monsanto's genetically engineered NewLeaf potatoes the following year. In consequence, Monsanto closed the research lab in Maine where the NewLeaf was developed and ultimately, pulled the potato from the market.

In Europe, these campaigns have succeeded in part because they tap into a smoldering anger about U. S. food imperialism.

This anger was graphically depicted in 1999 when activist farmers dumped four tons of GE grain in downtown London from a truck bearing a large sign that read "Tony, Don't Swallow Bill [Clinton]'s Seed. " When the French farmer José Bové, who trashed several McDonald's restaurants in his country and became a kind of folk hero, was asked what motivated his extreme actions, he replied, "To fight against globalization and advance the right of people to eat as they see fit". In the United States, activists have gathered support for their corporate campaigns by appealing to people's health concerns. One poster printed up by Greenpeace depicts a Kellogg's cereal box, with a caption reading "Kellogg's Frosted Fakes." The Kellogg's mascot Tony the Tiger also appears, saying "Frankenfood BAD!" In general, activists have portrayed the industry as greedy and irresponsible, more concerned with profits than with public health, the environment, or poor countries' food security.

U. S. markets in Europe, Japan, and other countries have been significantly affected by the negative reaction to GEOs in the food supply. Between crop years 1997–98 and 1998–99, for instance, European imports of U. S. corn fell precipitously, from 2 million to 137,000 tons. Worried about losing their export markets, some farmers are adopting a more cautious attitude towards new GE crop varieties, such as GE wheat. Gary Goldberg, CEO of the American Corn Growers Association, argues that farmers' reactions are motivated by their economic concerns: "For agricultural producers, this debate over GEOs is not a safety, environmental or health issue. It is an economic issue.... Simply put, can farmers afford to grow a crop that they may not have a market for?" These pressures have made it difficult for biotechnology firms to proceed with business as usual.

Evidence that the biotechnology industry has been hurt is apparent both in the defensive moves the big firms have made in the last few years and in the industry's all-out effort to redeem itself in the eyes of the public, its direct customers (farmers), and its investors. For years, the industry kept a low profile, quietiy developing and commercializing its new gene technologies, but that strategy has become impossible.The industry effectively admitted as much in November 1999, when seven of the largest life sciences firms formed a new industry alliance, the Council for Biotechnology Information (CBI), to counter criticism from antibiotechnology organizations. "The protest industry has gone too far," Edward Shonsey, a chief executive at Novartis Seed, told a *New York Times* reporter in November 1999. "They've crossed the boundaries of reasonableness, and now it's up to us to protect and defend biotechnology".

Through the CBI, the industry launched a $50 million advertising campaign to win back control of the discourse about agricultural biotechnology. Running the ad campaign are several high-powered public relations firms.

Individually and collectively, the life sciences firms are seeking to depict themselves as firmly committed to improving the social condition. One DuPont television commercial identifies one of its future tasks as "finding food to fight breast cancer. " Another portrays a poor old woman carrying a cooking pot in an unnamed Southern country; the caption proclaims that biotechnology will be able to "make food grow where food can't". The health and nutrition page of the CBI's Web site highlights four areas in which the industry is poised to improve the human condition: by producing foods that will deliver oral vaccines, reduce or eliminate allergies, be more nutritious, and last longer and taste better. In addition, the life sciences firms are undertaking highly publicized goodwill efforts in the battle to remake their image.

In early April 2000, Monsanto announced that it had a complete sequencing of the first crop-plant genome (rice) and then made this information publicly available by donating it to the International Rice Genome Sequencing Project, a consortium of ten publicly funded genome centres based in Japan. A month later, AstraZeneca announced that it would help make a genetically engineered strain of "golden rice" freely available to the developing world. But perhaps the most significant evidence that social activism and consumer concerns are hurting the biotechnology industry financially is the shake-ups that have occurred in the biotechnology stock market and the fortunes of individual companies. Monsanto's stock price fell by half in 1999 because of the company's troubles in the agricultural biotechnology arena. In March 2000, Monsanto was compelled to merge with Pharmacia and Upjohn, which then hived off the agricultural part of its operation and put it on the market.

Before it was sold, however, the new Monsanto Co. was radically restructured, its agricultural research budget was cut by 15 percent and narrowed to work on the four major crops, and the business was subjected to a stringent cost-cutting regime. As one Monsanto employee put it, "It went from a time when everything was possible and there were no limits to one where priorities matter and costs matter. It's a very different place". Another one of the "big five" companies in the industry, Aventis, sold off its agricultural division, Aventis CropScience, to Bayer AG, a move that was presumably related to its debacle with StarLink corn.

THE POWER OF ACTIVISM

In recent years, the literature on agro-food system restructuring has made much of the growing power of transnational corporations to determine the future of world agriculture. As states and firms have integrated themselves ever more tightly into the global economy and as private corporations have grown in size and economic power, large agribusiness firms have indeed augmented their reach and control over the agro-food system. But the power exerted by these transnational corporations is not absolute, unitary, or uncontested. In fact, a significant social movement has arisen to challenge this power and to try to alter the course of agricultural industrialization.

The antibiotechnology movement has certainly not stopped the biotechnology train in its tracks. But it has reduced its velocity, possibly altered its trajectory, and created a great deal of uncertainty for the life sciences firms by means of a vigourous and sustained political engagement with both the industry and governing agencies. Antibiotechnology activists have challenged the dominant discourse surrounding the promise of agricultural biotechnology and have offered an alternative framing of its potential social, economic, and environmental effects. They have tried to make states more accountable to their citizens by questioning the adequacy of national regulatory policy. They have supported supranational regulatory regimes that strengthen the hand of governments desiring to regulate imports of GEOs. And they have altered the economic calculus of the technology's developers.

Yet the power of the antibiotechnology movement should not be overestimated. Antibiotech activism has stimulated the emergence of a significant countermovement comprised of biotechnology firms, industry associations, and other proponents of the technology, including scientists, NGOs, and other non-profit institutions. Depending on the strategic behaviour of this countermovement and how influential it becomes, the impact of antibiotech movement activism may be attenuated.

The movement has far more limited access to financial, political, and public relations resources than do its adversaries. In addition, it is operating in a political context in which democratic participation in social issues is at an all-time low; and other political issues are considered more pressing than the food issue. Furthermore, the discourses of neo-liberalism, deregulation, and market competition now carry the day in policymaking. States and politicians feel a pressing need to support industries and undertake policies that foster national competitiveness in a rapidly globalizing economy. High-technology industries such as biotechnology clearly fall into this category.

Precisely who the movement represents and what democratic promise it holds are questions that require closer attention. Although we firmly believe that this new social movement has enhanced democracy by publicizing the biotechnology issue and bringing critical new voices into the biotechnology debate, it still behooves us to ask what the movement's relationship is to the larger societies in which it is embedded. Does it stand to be dominated by a few large organizations pursuing their own agendas, or will it remain diverse and participatory and closely connected to its grassroots?

Does the movement mainly address the concerns and fears of a group of highly educated, upper-middle-class, mainly white consumers, or does it reflect the interests of a broad range of citizens, including those for whom basic food security and the cost of food may rank much higher than preserving the right not to buy GE food? Indeed, is the movement committed to addressing the concerns of those whose primary worry is where their family's next meal is coming from rather than whether that meal is GE-free?

These questions cannot be answered here, but it is clear that they need to be addressed seriously, both by the movement's leadership and by its silent supporters. Finally, we need to determine the extent to which the movement can challenge the highly unequal power relations, both within specific countries and between them, that characterize the current agro-food system. This question of course focuses our attention on the movement's tactics and strategies, their potential for producing change, and their limitations.

A good example, as Julie Guthman suggests in this volume, is the strategy of supporting labeling, which can have the perverse effect (from the viewpoint of the activists) of legitimizing the use of agricultural biotechnology. Although such tactical and strategic decisions necessarily reflect the political realities of the environment in which any social movement operates (as well as its strategic resources and capacity), these choices invariably shape and constrain the political impact of social movement organization. The antibiotechnology movement at the beginning of the millennium is no exception.

Regulating beyond the Nation-State

Activists have also sought to slow the biotechnology train by helping to create supranational regulatory regimes that govern the international deployment of GEOs in food and agriculture production. This strategy is reflected in the movement's efforts to create an international protocol (known as the Cartagena Protocol on Biosafety) regulating trade in transgenic organisms, as well as in its ongoing efforts to influence the work of the Codex Alimentarius Commission, the UN organization that sets international standards for food health and safety. The idea for an international biosafety protocol was born during the negotiation of the UN Convention on Biological Diversity in 1992. During these negotiations, participants from the global South pointed out that their rich stores of biological diversity were uniquely vulnerable to the introduction of genetically engineered organisms. For this reason, Southern participants saw an urgent need for a protocol to oversee international trade in GEOs, which threatened to increase dramatically in the coming years.

But the United States and a handful of its allies (known collectively as the Miami Group) perceived such a protocol as a threat to the "right" of Northern farmers and corporations to produce and sell their goods, including genetically engineered agricultural commodities, worldwide and without prejudice. The group thus did its best to prevent the negotiation of such an agreement.

Ultimately, the powerful Miami Group lost. Although the final agreement was not as strong or comprehensive as most of the South wanted, it was broadly perceived by Southern negotiators and the antibiotech movement as a victory, since both groups had been seeking at least some form of control over GEO trade. Moreover, the fact that the regulatory regime would operate under the auspices of the relatively inclusive and democratic UN was perceived as a significant achievement, as the Miami Group had made a pre-emptive move to shift the onus of regulation to the clearly undemocratic World Trade Organization (WTO).

Although many factors contributed to the successful negotiation of the Cartagena Protocol in February 2000, the role of social activists was crucial. Prior to the negotiations, European activists and publics had put tremendous pressure on their governments to take genetic engineering seriously and to adhere to the "precautionary principle". This pressure induced the European Union to dispatch a number of high-level ministers to the negotiations, sending a clear signal to other parties (including the United States) that Europe intended to be responsive to its citizens.

Representatives from a wide range of NGOs, from the North and South, also attended the meetings in person and used a variety of tactics to pressure delegates to reach a meaningful agreement. The NGOs engaged in a division of labour, sending some members into the meetings as formal observers and spokespersons and others onto the streets to demonstrate. Activist groups organized all-night vigils in the freezing Montreal weather, trotting out one colourful image after another to keep the media's attention.

Activists on the inside also worked around the clock, lobbying delegates and disseminating up-to-the-minute economic, legal, and scientific information. Shortly after the negotiations began, the activist network released a position statement, signed by twenty-five influential NGOs from around the world, that demanded a strong biosafety protocol and admonished the Miami Group for obstructing the negotiations. Activist organizations also sought to shore up the power of Southern delegations.

Once it became clear that the United States had no interest in listening to the activists' position, they focused their efforts on forming a transnational alliance with state actors from the main negotiating bloc for the South, known as the LikeMinded Group.

The activists supplied these Southern delegates with information, helped them assess proposed text, and made concrete recommendations about accepting, rejecting, or changing specific language. They also brought environmental scientists to the meetings to give public presentations, and invited the delegates to attend. Reflecting on the negotiations, Tewolde Egziabher, the chief negotiator from the South, highlighted the importance of the activists' support when he explained how the Like-Minded Group, composed largely of African countries, managed to achieve anything against foes as powerful as the United States and its allies. "We had friends," Egziabher wrote.

Africa is financially so poor that the African Group would not have functioned ... without friends. But we soon made friends who filled in our gaps.... the Third World Network gave us critically needed assistance, and facilitated critically needed interactions, both South-South and South-North. African telecommunications are so poor that had it not been for the Gaia Foundation of London acting as an information relay station, we could not have been effective. And had our many, many other friends all over the world not helped, we would not have managed to stay as informed and as effective as we did.

As one observer put it, the activists "were the eyes watching for society at large, calling delegates on the carpet if they didn't remember that they have citizens back home who might not be happy if they do certain things.... They were the watchdogs." Many of the same organizations are playing a similar watchdog role in the context of the Codex Alimentarius Commission. In 1999, the *Ad Hoc* Intergovernmental Task Force on Foods Derived from Biotechnology was set up within the Codex to develop standards and guidelines for foods derived from biotechnology.

The definition of these standards and guidelines—which include risk-assessment criteria and labeling requirements and which recognize the role of the precautionary principle in food safety standard setting—lies at the heart of the political struggles over the task force's work. Activists have followed the committee's deliberations closely, studying and critiquing the official positions taken by governments and their representatives and pressuring governments and the task force itself to try to influence its decisions.

A central goal of these activist groups is to ensure that task force decisions do not contradict or override the biosafety protocol's provisions. Activists are of course not the only ones trying to affect these decisions;

every group with a strong interest in the Codex's decisions has sought to get its voice heard. In fact, one could argue that the activists' efforts are simply aimed at counteracting the influence of far more powerful Northern agribusiness interests.

Generally speaking, however, activists' efforts to establish and shape new international regulatory regimes for genetically engineered foodstuffs have helped to make decision making about agricultural biotechnology more democratic by ensuring that decisions about the deployment of biotechnology are not left entirely up to those with the most to gain from promoting these products and the least to lose if the products have negative economic or environmental consequences in importing nations.

3

Agri-Biotechnology and Agro-Food System

NEW PERSPECTIVES ON AGRO-TECHNOSCIENCE

Although the labour process-commodification perspective retains analytical appeal, it is clearly a product of its times. As Kloppenburg observes, "The model of change that emerges from this analysis is fundamentally dialectical—the forces and relations of production are mutually conditioning". Human praxis is ontologically central, and nature is firmly located on the opposite side of the modernist divide.

At best, nature is endowed with certain qualities of "resistance," which are identified, for example, with "biological barriers" to capital and "secrets of life" that are inaccessible to science. However, in this characterization, nature is figured as a passive entity whose latent properties will be revealed only if it is *acted upon* by industrial technoscience. Following the Enlightenment antinomy between nature and society, nature is other, which industrial society is driven to subjugate and appropriate.

As depicted in *From Farming to Biotechnology*, the agricultural labour process has indeed presented historically contingent biophysical limits to industrial appropriation and substitution. Yet, although nature and industry are drawn as oppositional categories, in analytical terms, industry is the only actor. This unexamined ontological choice precludes the conceptualization of nature as a lively, active, and formative presence and so *underplays* what arguably are the truly revolutionary material dimensions of biotechnology.

Unwittingly and implicitly, with this choice the analysis is captured by the *engineering* metaphor of life propagated by industrial capital, with its omnipotent technocratic discourse of precision, efficiency, and benign evolutionary improvement.

This observation is not intended to deny the incisiveness of the labour process framework as an analytical vantage point to address social *cui bono* issues and rural structural change as nature is reconfigured as a productive force. Clearly, such a denial would fly in the face of historical and contemporary evidence of the salience of production as a terrain of theory and praxis. Nevertheless, the ontological critique of late post-industrial capitalist political economy can fruitfully be extended by bridging the modernist divide between nature and society in order to imagine and construct alternative socioecological worlds. As already noted, agency in "first generation" approaches to ABTs is uniquely identified with human intentionality and human action on an objectified, but now less recalcitrant, nature. This instrumentalist ontology, with its purified categories of nature and society, is closed to notions of the relational materiality of nature, offers no theorization of the lively entities emerging as social partners in these technoscientific practices, and fails to entertain nonhuman perspectives and shared consequences.

An alternative conceptualization of nature, one that attributes active properties to nonhuman entities, would focus analytical attention on the interrelational negotiation of new socionatural realities, which are now being constituted with the commercial diffusion of genetically engineered crops in the United States. Such an approach also would make explicit the ethical, environmental, and political choices associated with different human-nonhuman assemblages and "socio-ecological projects". However, as argued elsewhere, mainstream agro-food studies remains transfixed by the modernist ontology. Encouragingly, there are recent signs of a reflexive "turn" to give analytical salience to "the status of nature" and to interrogate the modernist divide.

Agro-Social Networks Theory

The analytical richness and potential of actor-network theory (ANT) in agro-food studies have been expressively revealed in these reflexive explorations. However, apart from an illustrative vignette, these analytical resources so far have not been deployed in work on ABTs. This is a major omission in agro-food studies, and efforts to rectify this position deserve high priority on the ABT research agenda.

Lawrence Busch and his colleagues in a recent series of papers on the rapeseed (canola) commodity sector have undertaken the most systematic work on agricultural technoscience from an actor-network perspective. This impressive research elucidates the endogeneity of technical change and the processes by which knowledge-commodity transformations become ecosocially embedded.

ANT is used to analyse the rise and institutionalization of rapeseed technoscience by conceptualizing technological innovation as a process of network building. Thus, instead of considering technology as an exogenous factor to the system of commodity production, the network approach does not differentiate between human and nonhuman elements. In such analyses, society does not transform technology, nor does technology cause the transformation of society. Rather, "the very actor-networks ... simultaneously give rise to society and to technology." Moreover, Arunas Juska and Lawrence Busch find ANT to be particularly helpful in uncovering points both of analytical entry and of political action in cases where, as with ABTs, processes of network formation, extension, and reconfiguration are dynamic and in flux. In this perspective, "it becomes evident that the relationships mediated through the network are contingent in nature: they can be disrupted; they can collapse; they can be organized according to different principles; they can be constantly changed and renegotiated".

In a second paper, Busch and Keiko Tanaka (1996) deploy the concept of symmetry, which ANT extends to nonhumans, to give "an explicit portrayal of non-human actors over time and space" in "the complex networks known ... as commodity chains", such as rapeseed (canola). The product grades and standards constitutive of commodities represent "rites of passage" for both nature and people. "Thus, by transforming nonhumans and subjecting them to multiple rites of passage, we coproduce nature, society, [and] the capitalist market". The symmetry of these qualification processes also extends to their ethical dimensions, which infuse these modes or "rites" of mutual socialization of humans and nonhumans. In a later analysis of the globalization of rapeseed networks, Busch and Juska contend that the political economy of agriculture perspective and its "categorical apparatus" are "inadequate" to the task at hand.

The political economy approach is underspecified because "in true Baconian style, nature is recast as resources to be transformed", as passive. In contrast, the political economy of actor networks reveals how the strategic positions of human and nonhuman actors and the geometry of power in networks are negotiated, modified, and transformed temporally and spatially. A more recent paper investigates these questions in finer historical detail by examining the reciprocal and contingent relationships between agricultural research and agricultural production, and the changing strategic positions of nation-states as the rapeseed sector has become more globalized since the mid-1970s.

This brief review of rapeseed technoscience provides some indication of the purview and analytical purchase of ANT. With its symmetrical ontology and method, ANT offers a penetrating conceptual repertoire for the analysis of ABTs. Lively eco-social coproductions and category fusions of the natural and artifactual are particularly in evidence in the everyday practices of agro-food networks in an era of genetically engineered organisms. Thus Bruno Latour extends an invitation to reject the purified categories of "society" and "nature" and focus instead on "the blind spot where society and matter exchange properties". These transactions are brought centre stage in order to expose the inescapable sociomateriality of the entities mobilized into the heterogeneous associations that hold society together. The constant interchange of human and nonhuman properties in network formation has created "mixtures between two entirely new types of being, hybrids of nature and culture". These processes of mediation, in turn, constitute the foundation of modern technosciences, which "multiply the non-humans enrolled in the manufacturing of collectives and ... make the community that we form with these beings a more intimate one". Rather than engage in a lengthy exposition of ANT, its merits and limitations, the point to emphasize is that we have a nondualist, relational, and processual framework in which nonhumans are actively present, performative, and consequential. In this significant respect, although otherwise confessedly modest in its claims, ANT challenges the silences and abstractions of production-centreed analyses.

That is, it furnishes an ontology and conceptual language with which to address the implosions, natureculture hybrids, and newly socialized intermediaries emerging from the heterogeneous engineering practices of agro-food technoscience. ANT's terms of engagement and insistence on "the permeable boundary running between humans and non-humans" resonate strongly with the ethical and relational concerns of biopolitical activism. An ethical standpoint, a relational moral philosophy, is discernible in ANT's framing of the construction of our world, its insistent reference to crossovers and the exchange of properties between human and nonhuman entities. As Jonathan Murdoch notes, actor-network theorists "force us to look afresh at the categories, divisions and boundaries that frequently divert our attention away from the nonhuman multitudes which make up our world".

This imperative recalls Latour's analysis of Louis Pasteur's research on the anthrax bacillus. That is, the bacilli, once "translated" from the natural competition found in the farmyard to Pasteur's laboratory in Paris, encounter a new environment, an altered state, where they can thrive. Following Latour, Elizabeth Bird observes that "In the laboratory, nature—in the form of Pasteur's "natural-technical object"—becomes an actor negotiating a *new reality*. In the terms of that context, the microbes become actors in shaping a new environment" (258, original emphasis). Latour continues, "Training microbes and domesticating them is a craft.... Once these skills have accumulated inside laboratories, many cross-overs occur that had no reason to occur anywhere else before".

Similarly, when analyzing Pasteur's discovery of the microorganism responsible for lactic fermentation, Latour draws attention to the change in ontological status that this step involved and observes that "in his laboratory in Lille Pasteur is *designing* an *actor*". The analytical parallels and possibilities of extending this approach to the transgenic "natural-technical objects" of ABTs—such as Roundup Ready soybeans and Bt corn, not to mention the unintended progeny of horizontal gene transfer—are clear. In a brief vignette on ANT, I suggest that agri-biotechnological innovation can be analysed in terms of the practices deployed to reconfigure or "translate" existing agro-food networks by enrolling genetically engineered organisms (GEOs) and foods. In effect, corporate networks, such as Aventis, DuPont, and Syngenta, are seeking to displace previous socionatural orderings and to realign agro-food networks in ways that support and "naturalize" the diffusion of GEOs into rural environments, crops, and animals.

agri-biotechnologies introduce new mediators

Thus, "agri-biotechnologies introduce new mediators into the intimate corporeal relations of agro-food networks, promising new corporealities and, quite literally, new bodies". Biopolitical mobilization, notably in Western Europe, has raised material and ethical concerns against this reordering of agro-food networks in marshaling opposition to the environmental release of GEOs and the incorporation of genetically engineered foods into human bodies.

This mobilization directly challenges the industrial, technoscientific problematization of ABTs, variously framed as the answer to world hunger, as an improvement on nature, or in terms of the inevitability of technical change. In this modernist instrumental perspective, nature is objectified as a field of resources awaiting domination and exploitation by the relentless advance of technoscience. Against this dualist rationality, green biopolitical activism is informed by a relational ethics and by precepts of shared community.

These biopolitics of ABTs reveal the "clash of divergent ontologies" provoked by struggles to realign agro-food networks into new socionatural orderings. By undermining the modernist ideology of nature as externalized and objectified, ANT provides theoretical resources to address nature and its lively materiality directly. This attention to how "socionatures" are constructed broadens critical engagement with capitalist technoscience and political economy and informs our understanding of the heterogeneous associations fostered by this ordering of the socioecological.

The dimensions of this political space are amplified if we interpret ANT as an ethical discourse of how to live in the world. This perception of ethical, as well as material, embeddedness speaks directly to the problematics of biopolitical activism, as noted above. Even a cursory understanding of ABTs brings recognition of the new socionatural assemblages emerging, under the aegis of capitalist technoscience, to construct new worlds. In turn, this recognition suggests the need for conceptual frameworks that explicitly bring normative judgement and political critique to bear on these new collectives and respond to Donna Haraway's interrogation of technoscience: how, for whom, and at what cost? Acknowledgment of our partnership in these human-nonhuman assemblages would be an initial step towards the development of forms of social organization that encourage democratic choice between alternative orderings and the worlds they bring forth.

For these purposes, we need ontologies that reveal, not abstract, our interactive, relational production of worlds we inhabit with others. The novel socionatural assemblages of capitalist agro-food technoscience, which herald new actors and new realities, underline the significance of theoretical and political choice. In this re-encounter with agro-food studies research on ABTs, the theoretical and thematic continuities of this literature emerge insistently. Indeed, in theoretical terms, there is a powerful sense of deja vu, of involution even, and alternative perspectives remain firmly on the margins. I emphasized both the primacy of the labour process commodification approach and its ontological limitations for comprehending the new socionatures constituted by ABTs.

This framework constrains efforts to find common ground with the relational ontology and moral economy that inform the biopolitics of environmental movements and Green activists. Such serious limitations, in short, lend urgency to theoretical renewal in agro-food studies and, more generally, to the "greening" projects and explorations of Red-Green rapprochement now under way in various fields of critical social theory. These projects unmask the political agendas and instrumentalist ethics imbricated in modernist ontological antinomies and their reification. At the risk of repetition, the agro-food studies literature on ABTs has as its analytical focus the ensemble of institutional processes and social relations that have led to the commodification of nature or, as Yoxen (1981) has it, to "capitalizing life".

This labour process conceptualization, with its emphasis on the subsumption of nature and its manipulation as a productive force, sees this transition exclusively from the standpoint of the social. Nature is subsumed by purposive social agency, whose dynamic is to be found in the laws of accumulation and social relations of the capitalist mode of production. There is no place here for the relational materiality of nature, its liveliness, or its "boomerang" qualities.

This framework does not entertain either notions of natural-social coproductions or the consequences of these assemblages for entities with whom we share this world. The labour process perspective, in short, does not lend itself to an assured engagement with the new constituencies of agro-food biopolitics. However, "this is not to dismiss the strengths of this perspective. Rather, it is ... to observe that this theoretical lens or 'framing' device does not focus directly, for example, on the new socionatural relations, interspecies metabolisms and exotic corporealities unleashed by agricultural biotechnologies.... these new constellations or assemblages of nature-society relations are key catalytic elements of bio-political activism in agro-food networks". This review also has drawn attention to thematic continuities, although clearly the scale and intensity of social mobilization have grown sharply with the accelerating deployment of ABTs. These thematic links are amply reflected in this volume. Issues of governance, although not known as such in the 1980s, form a recurrent body of concern. Two themes, already identified previously, have retained particular salience. The first theme concerns the ways in which ABTs heighten the concentration of industrial control in agro-food systems—"a global oligopoly" in William Boyd's estimation. Exploring the "deep structures of monopoly" within a commodification of nature framework, Boyd extends the analysis of ABTs into the era of life sciences multinationals, genomics, and the competitive imperative to capture value by integrating vertically from proprietary intellectual-property platforms to seed marketing and contract farming.

A social constructivist-commodification perspective also frames the contributions of Scott Prudham and Dennis Kelso to this collection. Following Richard White, Prudham adopts the metaphor of nature as an organic machine to describe the transformation of living organisms into technologies and commodities. Within this framework, he is particularly concerned to track the harnessing of public science to private innovation in the development of forestry biotechnology, the concomitant restructuring of university-industry-state relations, and how this trajectory is shaped by the specificity of recalcitrant nature. In warning against technological fetishism, Prudham insists that biotechnologies be seen not as things but as bundles of social relations with historical lineages in order to emphasize that technological change is socially produced.

Although Kelso's analysis of the deployment of biotechnology in commercial salmon aquaculture broadly fits with the labour process-agrarian transition problematic of agro-food studies, his main concern is to reject its technological determinist inflections. This endeavor brings Kelso into closer engagement with issues that have been strangely muted in this literature until recently and especially with the sources, forms, and resources of social resistance to biotechnological innovation. Kelso's account of transgenie salmon embraces not only mobilization around questions of food safety and environmental risk but also the defensive stance of salmon farmers against the perceived technological threat to the aquaculture industry as presently constituted. In elaborating these questions, Kelso draws attention to the politics of scientific uncertainty and state regulation, highlighting the strategic importance of discursive struggles to form public perceptions of nature and the natural.

This discussion, together with the chapters by Julie Guthman, Frederick Buttel, and Rachel Schurman and William Munro, begins to address important lacunae in the agro-food studies literature. The second thematic grouping is around governance issues and focuses on continued First World-Third World tensions over access to genetic resources and the perceived asymmetries of power articulated by intellectual property rights regimes and, notably, the 1994 WTO TRIPS accord. These ongoing tensions find expression in the contested politics, shifting alliances, and arcane regulatory processes of new multilateral institutions of global governance—the WTO and CBD—and their disputed mandates in the conjoined policy arenas of international economics, trade, and environment. Environmental governance emerges in another contemporary guise in the chapter by Astrid Scholz.

Her analysis traces the vagaries of the utilitarian rationale for biodiversity conservation as the importance of natural product screening in the R&D strategies of transnational pharmaceutical corporations waxes and wanes, exacerbating the inequities of power embedded in private bioprospecting agreements. The contribution by Frederick Buttel similarly can be situated within the global governance thematic; however, like Kelso's chapter, it also provides a bridge to some neglected issues. Buttel's wide-ranging account encompasses the institutional architecture of the globalization regime, the power brokering of the GATT-WTO transition and the "spoiling" role of agro-food issues—festering EU-U. S. "food wars" (bananas, beef, GEOs), unilateral trade sanctions, the "environmentalization" of ABTs, the global farm crisis—in revealing potential fault lines and pressure points in this accumulation regime. Taken in the aggregate, these individually contentious issues have cumulatively resonated with growing force in political, cultural, and institutional domains. Whether or not GEOs prove to be the Achilles' heel of the globalization regime, Buttel is surely right to stress that EU-U. S. disputes, and their potential to galvanize social protest, are less about trade liberalization than about the perceived threat of institutional convergence, and especially the forced harmonization of national regulatory structures and the further erosion of cultural identity.

The bridge in Buttel's account rests on recognition of the crucial nexus formed by the politics of scientific research and policy, environmental risk, food safety, regulation, and consumption. These politics are complexly intertwined and varied in their manifestation, ranging from agro-food movements, environmental coalitions, and many forms of direct action, including street theater and green sabotage, to NGO involvement in regulatory processes and consumer pressure on food retailers and manufacturers. This terrain remains largely unexplored in agro-food studies, apart from limited incursions into one or other of its arenas, such as the politics of rBGH consumption or the role of ABTs in facilitating transition to demand-driven food systems.

This neglect is surprising in view of the considerable prominence of these research themes in other fields of critical social theory. In this context, work on the institutional matrix of science, science policy, and regulation, as undertaken more generally by Yoxen, Kay, Bud, and Wright, could fruitfully be extended to developments in agriculture and food since the 1980s. The chapters by Guthman and by Schurman and Munro set some markers to follow in addressing the research nexus identified above. Guthman examines the contradictions that GEO labeling and right-to-know legislation in the United States present for movements seeking to build an effective politics of consumption, especially the privatization of risk management for genetically engineered foods and the political disarticulation that labeling implies. In calling for research on "an emerging political economy of risk," Guthman follows Haraway (1997) in emphasizing the centrality of struggles for "the power to define what counts as technical or political". In this respect, it would be interesting to explore how European consumer groups and environmental organizations have avoided the pitfalls of labeling and convinced national governments and the EU to redraw the boundary between the technical and the political by agreeing to reopen previous regulatory decisions to public scrutiny.

Demarcation struggles and boundary changes are the central theme of Schurman and Munro's chapter. The chronology of events in the rising social opposition to ABTs is becoming well-known, but there is a dearth of careful analyses of antibiotech activism, its strategies and modus operandi. Schurman and Munro begin to address this hiatus by examining antibiotech movements in the North, primarily in the United States.

In their view, the marked shift in public sentiment against genetically engineered foods is not attributable to inchoate consumer resistance or to spontaneous protest. Rather, these dramatic changes, and its economic and regulatory repercussions, have been orchestrated by the social-movement organization and mobilization of established civil-society actors. In pressing this argument, the authors provide valuable insights into the wide array of social organizations articulated by this movement, its tactical sophistication, and its networking skills, which facilitate operation across a variety of institutional and spatial scales and in different regulatory spheres and discursive arenas. Schurman and Munro leave unanswered the thorny questions of representation and the transformative potential of the antibiotechnology movement. However, it is to be hoped that this initial exploration will encourage others to join them in investigating this serious gap in the politics of agricultural technoscience.

In December 2000, completion of the plant genome sequence of thale cress (*Arabidopsis thaliana*) was announced, the first of some 250,000 plant species. Agro-food technoscience, here represented by the public-private consortium, including Monsanto, which comprises the Multinational Coordinated *Arabidopsis thaliana* Genome Research Project, goes marching on. Now, as in the 1980s, it is vital that critical social scientists join with the antibiotech movement and other progressive forces in the struggle to submit this enterprise and the worlds it would create to democratic debate and public assent.

new technoscientific paradigm in Agri-biotechnolgy

Although ABTs are represented as a new technoscientific paradigm, the analysis is anchored in the agricultural labour process and its transformation via commodification of seed production and plant breeding. In an italicized passage, Jack Kloppenburg stresses that "the seed, as embodied information, becomes the nexus of control over the determination and shape of the entire crop production process."

The wider structural implications of ABTs in terms of industrial appropriationism and substitutionism, or institutional change and new property forms, for example, are similarly ushered through the privileged gateway of production. This obligatory passage-point into the circuits of capital also is seen in the preoccupation with industrial concentration, reflecting the redistribution of power towards upstream farm-input sectors as agrochemical and pharmaceutical companies take up dominant positions in ABTs and the seed industry. Broadly speaking, this labour process-commodification perspective has provided the general, overarching analytical framework of choice for addressing ABTs in agro-food studies. The "first generation" analyses approached ABTs by problematizing the boundaries between agriculture and industry. This analytic move provided the basis for more systemic and historically informed studies of agroindustrial development by drawing attention to the contingent nature of this division of labour. Nevertheless, the farm labour-process is taken as the privileged locus of the transformative contradictions generated by agrotechnoscientific innovation.

The work of Kloppenburg, for example, is first situated generally on the classical Marxist terrain of agriculture as a "recalcitrant sector" and then takes up the specific theme of the vectors of capitalist penetration of public plant-science and its gradual institutional reconfiguration as "capitalist property". The commodification of the seed is conceptualized as a process of primitive accumulation, a process whose highlights are the innovation of hybridization—which "functions to uncouple agricultural producers from the autonomous reconstitution of their own means of production"—politically driven shifts in the institutional division of labour between public and private plantbreeding, and changes in intellectual-property regimes to facilitate the private appropriation of plant genetic resources.

Kloppenburg's purpose is to explain how the seed, "which is perhaps the element of agricultural means of production most central to the entire farm production process", is commodified and becomes a capitalist force of production. Agricultural biotechnologies are emblematic of this historical trajectory because "what is now occurring in the seed sector is one instance of a much broader technological transformation that is galvanizing changes in the social organization of all production processes in which organic substances or life forms play a significant role". My colleagues and I apply the commodification approach to on-farm means of production more widely (implements, motive power, nutrient cycles, pest control, seeds, energy) and argue that the biophysical processes of agricultural production and food consumption have constituted natural, though historically contingent, constraints to the industrialization of agricultural use-values.

Unable to reproduce natural production processes fully by direct transformation, industrial capitals have adapted in singular ways to the sectorally differentiated properties of agricultural nature (biological time, photosynthesis, land, climate) and the physiology of human nutrition. These differences, we argue, are analytically significant as a major source of variation in the historical dynamics, specificities, and contemporary configurations of social production in agro-food commodity networks. The concept of appropriationism is used to designate the historically discontinuous, piecemeal "but persistent undermining of discrete elements of the agricultural production process, their transformation into industrial activities, and their re-incorporation into agriculture as inputs". Elements of natural production processes are progressively internalized by industrial capitals via proprietary science and technology as individual sectors of capitalist accumulation and reproduction.

In brief, in agriculture, where industrial capitals confront a natural production process, agricultural biotechnologies constitute "a generalized advance in the capacity of industrial capitals to manipulate nature". Whether in explicit or implicit terms, the framework adopted by my colleagues and I and by Kloppenburg (1988) posits a contingent, ongoing *tension* between agriculture as a recalcitrant sector and its full assimilation by industrial capital. Though with differing emphases, these studies broadly explore the cumulative effects and convergence of several interdependent processes of incorporation—cognitive, technoscientific, economic, and regulatory—that progressively extend the commodity form to new spheres of the farm laborprocess. In this context, the technoscientific paradigm constituted by biotechnology is identified as the catalyst of assimilation.

Nature is transmuted into a force of production. This reconfiguration represents the vector of "domestication" of recalcitrant biological processes, hitherto inaccessible to technoscientific manipulation and the reductionist endeavors of industrial capitals. In this framework, biotechnology, actually or potentially, has swept away the *technological* foundations of the recalcitrance or exceptionalism of agriculture. For example, we formulated the basic question of the agricultural labour process-commodification approach in the following terms: "If biotechnology represents a qualitative breakthrough in that nature can now be reconstituted industrially, does that mean that the food system is open to assimilation within the broader transformations of the industrial system?". The answer we gave at that time could not be clearer; the dichotomous tension is sundered definitively. "In this perspective, biotechnology ... markfs the end of the pre-history of the food industry and its incorporation within the broader dynamics of the industrial system and post-industrial society".

However, with what now seems fortunate prescience in the light of contemporary biopolitical activism, we suggested that "The frontiers of substitutionism are likely to be defined as much by consumer tastes and loyalty to organic whole foods as technical and engineering constraints".

commodification approach in Vectors of Incorporation

The labour process-commodification approach clearly has furnished the preferred analytical framework for research on ABTs in agro-food studies. A recent critique and extension of "first generation" analyses attests to the continuing influence of this approach. This same theoretical perspective also frames a number of subsidiary research themes, which emerged in the 1980s and early 1990s as social scientists came to grips with the wider implications of ABTs. A full account of these complementary studies, whose scope ranges from the institutional intricacies of regulatory change to the ethics of genetic engineering, exceeds the more limited purview of this essay.

However, with some qualifications, these subsidiary studies follow the same analytical threads Kloppenburg (1988) traces in his history of plant improvement: that is, "the commodification of the seed, the changing division of labour between public and private research institutions, and the appropriation of plant genetic resources". A particularly rich vein of scholarship on ABTs is devoted to the constellation of political-economic forces that determine the shifting demarcation line between basic and applied research, and the corresponding realignment of the institutional division of labour between public- and private-sector research.

The changing division of labour in the U. S. agricultural research establishment, and notably the institutional "capture" of the land-grant university system, dominated work on ABTs in the 1980s. These studies grappled with the political-economic and ideological issues raised by the mounting evidence of corporate penetration of American research universities, where "biotechnology was born". As Edward Yoxen puts it, the capitalist incorporation of molecular biology has reached the point where "the industrial exploitation of recombinant DNA research by corporate capital can be serviced directly from the academic research laboratory".

With remarkably few exceptions, the agro-food studies literature has been reluctant to follow Yoxen's lead and venture seriously into the history of science. However, the emergence of the "universityindustrial complex" and the vital contribution of academic scientists to the nascent biotechnology industry are explored fully in Martin Kenney's outstanding and prescient book *Biotechnology: The University-Industrial Complex*. This study crowned a wave of related papers investigating the emerging social division of labour in agricultural biotechnology and plant-breeding research. Much of this work on institutional change is associated with Fred Buttel and his colleagues, then in the Department of Rural Sociology at Cornell University. The rapidly changing matrix of agricultural research policy, scientific practices, and public research institutions and the international reach of agricultural biotechnologies also distinguish the scholarship of Lawrence Busch, Bill Lacy, and their collaborators in this same period.

A second strong theme of agro-food studies research on ABTs in the 1980s focused on patterns of innovation and industrial concentration, and their impacts on agricultural production and rural social structures. In a pioneering paper, Jack Kloppenburg (1984) assesses the prospective structural consequences of ABTs by extending the historical tendencies already observed in the development and diffusion of hybrid corn. As in this earlier case, "Biotechnology, too, promises to create a vast new space for the accumulation, concentration, and centralization of capital".

This Schumpeterian notion is reinforced by other analytical foci, including the capitalist subordination of public agricultural research and plant breeding, the development of herbicide-resistant crops, the loss of farmer autonomy, acceleration of the technological treadmill, and continued genetic erosion. If the concentration of intellectual-property ownership in corporate hands is added to these foci, a remarkable continuity emerges between the scholarship of the early 1980s and the contemporary political-economic analysis of ABTs. A further parallel can also be drawn with current activist opposition to the rising corporate control of global agro-food systems, opposition that has captured public attention since the street battles at the 1999 WTO meetings in Seattle. Other contributors approached the socioeconomic impacts of ABTs by examining innovation patterns, typically through the prism of new opportunities for accumulation, the implosion of sectoral barriers to entry, and the industrial reorganization and concentration in the agro-food system as a whole.

These studies focus on the upstream and downstream industrial sectors, notably the chemical-pharmaceutical complex, to draw out the implications of ABT innovation for farming and for social-production relations in farming. For example, John Wilkinson and I suggest that the generic capacity to engineer living organisms "prefigures a new bio-industrial paradigm". "Biotechnologies now threaten to *implode* the long-standing organization of the food system around specialized commodity chains....

There is the capacity simultaneously to relocate agricultural production in factories and industrial production in fields". Although countervailing tendencies are likely to set up tensions between alternative models and trajectories, we argue that ABTs will accentuate the shift towards a more demand-oriented agro-food system. Biotechnology as the generic catalyst of wider impulses towards bioindustrialization also underpins projected scenarios of change in the social organization of agriculture. These processes include the amplification of the observed trend towards large-scale, intensive, continuous production systems and the introduction of all-purpose agricultural "biomass refineries." Such developments are predicted to lead to greater concentration of land ownership and more widespread contract production and part-time farming.

From the early 1980s, innovation studies with a more action-oriented policy research perspective began to appear in the agro-food studies literature. These studies were galvanized by the field testing and approaching commercial adoption of genetically engineered organisms, which gave concrete expression to concerns about environmental safety, rural structural change, and demands for regulatory processes with greater democratic participation. Struggles over the licensing of recombinant bovine growth hormone (rBGH) crystallized these issues acutely in the United States. The rBGH controversy also exposed the conflicted politics of ABT research in the U. S. land-grant university system and other public institutions involved in the innovation process. A third and continuing strand of agro-food studies scholarship on ABTs examines the political-economic repercussions and contested politics of changing intellectual-property regimes in plant breeding and genetic resources.

As in the case of rBGH, the academic literature in the 1980s was strongly informed, if not led, by parallel activist contributions, notably from members of RAFI, or the Rural Advancement Foundation International. These interactions have intensified since the 1992 Earth Summit in Rio as part of the ongoing struggles to shape the emerging supranational institutions of global environmental governance, and notably in this context, the Convention on Biological Diversity (CBD) and the Cartagena Protocol on Biosafety, which regulates international commerce in living modified organisms. Contributors in the 1980s quickly grasped the import of the landmark rulings that extended and consolidated property rights to microorganisms, plant germplasm, and rDNA techniques and other processes involved in plant genetic research. The dramatic institutional shift to allow the cornmodification of life forms—"ownable artifacts, whilst also being a part of nature" — has been analysed from various angles in agrofood studies.

Several accounts trace the institutional origins and evolving structures of intellectual property rights in plant breeding before 1980, when the U. S. Supreme Court's decision in *Diamond v. Chakrabarty* overturned the "product of nature" principle, which had earlier dominated this field of patent law. Other scholars surveyed this same terrain but from specific institutional standpoints to evaluate the implications of these new legal and social conventions for particular groups of actors. This scholarship, for example, investigated university-industry relations and the public and private communities of agricultural scientists and research administrators. However, in agro-food studies, the issues arising from the patenting of life forms were explored most vigourously at the international level. Academics and activists exposed the institutional dimensions of NorthSouth power asymmetries and the inequalities embedded in the governance structures of international agricultural research, such as the Consultative Group on International Agricultural Research (CGIAR) system, and the associated collection and transfer of germplasm as "common heritage" resources. Jack Kloppenburg deserves much credit for bringing the "North-South seed wars" and questions of political-economic control over international plant genetic resources into agro-food studies, and other scholars soon followed his lead.

More recently, issues of access to and ownership and conservation of international genetic resources have faded from prominence in this literature. This decline, paradoxically, has coincided with the rising salience of these questions on national and international policy agendas as Greenpeace, Friends of the Earth, and other leading environmental groups have reinforced the longstanding campaigns of more specialized, agrienvironmental organizations, such as RAFI and Genetic Resources Action International (GRAIN).

Although incomplete, this survey of the earlier literature on ABTs reveals some strong continuities with contemporary scholarship, in terms of both theoretical perspective and more specific concerns, such as genetic erosion, decline of farmer autonomy, and centralized control of agro-food networks on a global scale. Discursive continuities also persist, despite Fred Buttel's suggestion that the days of extravagant rhetoric characteristic of biotechnology in "its formative years, roughly the mid-1970s to late 1980s," are now past. In this period, Buttel argues, virtually all the protagonists—leading molecular biologists, venture capitalists, small startup firms, government officials, and environmental and public-interest groups—embraced it as epoch-making and revolutionary. "As late as the end of the 1980s, most academic, policy and activist treatments of biotechnology were essentially agreed on its magic bullet character and transformative potential".

At the turn of the millennium, the implication that this viewpoint is no longer shared seems premature, even if more pragmatic assessments can increasingly be found. The tropes of these formative years are still actively deployed in the rancorous exchanges between supporters and opponents of ABTs. For each group, the discursive starting point remains the promise of the technology: golden rice versus Frankenfoods, precision breeding versus superweeds, and so on. A brave new socionatural world is on offer, which, if realised, would empower the networks of human and nonhuman entities involved in its construction. Some recent work on technoscience and relational ethics in agro-food studies, particularly extensions of actor-network theory, seek to elucidate such biopolitical choices by adopting an explicitly nondualist framework of analysis. These new approaches represent a reflexive theoretical project to overcome the abstraction of nature in mainstream agro-food studies by interrogating its modernist ontological foundations.

The Future of Biotech Foods

Since the known benefits of biotech foods outweigh the uncertain risks, the precautionary principle supports the use of such foods. In the 1980s, the precautionary principle migrated from its European roots to the international environmental scene. By 1992, it was ubiquitous: it appeared in the Rio Declaration, the United Nations Framework Convention on

Climate Change, and the Convention on Biological Diversity (CBD). These precedents ultimately spawned the CBD's January 2000 Cartagena Protocol on Biosafety, which uses the principle as a basis for assessing the risk of international transfer, handling, and use of genetically modified organisms. This protocol is considered a major victory for the precautionary principle and its advocates.

A popular formulation of the precautionary principle is the Wingspread Declaration: "When an activity raises threats of harm to human health or the environment, precautionary measures should be taken even if some cause and effect relationships are not established scientifically." This principle has been interpreted by many to require—as precautionary measures—policies to curb technologies that science cannot prove to be absolutely safe. Specifically, it has been invoked to justify a ban on research, development, and deployment of genetically modified (GM) crops. This application of the precautionary principle acknowledges the public health and environmental risks that a ban might reduce but ignores other risks that a ban might generate or prolong. Therefore, while the operation might be successful, the patient might die.

Moreover, the principle provides no guidance in situations where actions such as a ban on GM crops could simultaneously generate uncertain benefits and uncertain harms. To avoid counterproductive "precautionary" policies, I propose a framework for applying the precautionary principle. This new approach provides a way to evaluate policies that might result in ambiguous outcomes because the harms might offset, in whole or in part, the benefits of GM crops. Using this framework, the precautionary principle can be used to answer a variety of difficult questions in the debate over GM crops. For example, do the public health and environmental benefits of GM crops exceed their harms? Considering the wide range of consequences of banning such crops, would the precautionary principle, in fact, justify such a ban? Finally, would a ban be consistent with the stated aims and objectives of the CBD and the Biosafety Protocol?

Competing Uncertainties

This framework consists of a set of hierarchical criteria to rank various threats based on their characteristics and likely consequences. The first criterion is human mortality: the threat of death to any human, no matter how humble, outweighs similar threats to members of another species, no matter how magnificent. Moreover, the other nonmortal threats related to human health—such as blindness, stunting, additional days of sickness, or lowered educational attainment—should take precedence over threats to the environment, although exceptions might be made based on the nature, severity, and extent of the threat. This is the human morbidity criterion. These two criteria can be combined into the public health criterion. Additional criteria must be considered if an action under consideration results in both potential benefits and harms to public health or the environment. These other criteria include:

- The immediacy criterion. All else being equal, more-immediate threats should be given priority over threats that could occur later. This criterion is justified because people tend to partially discount the value of lives that might be lost in the distant future.
- The uncertainty criterion. Threats of harm that are more certain should take precedence over those that are less certain if their consequences otherwise would be equivalent.
- The expectation value criterion. Other things being equal, precedence should be given to threats with higher expectation values. For example, actions resulting in fewer expected deaths should be preferred over those resulting in more expected deaths.
- The adaptation criterion. To the extent technologies are available to cope with, or adapt to, an impact, then that impact ought to be discounted.
- The irreversibility criterion. Greater priority should be given to outcomes that are irreversible, or likely to be more persistent.

Weighing the Benefits

Before making any decisions regarding new technologies, including GM crops, we first need to assess the harms and benefits associated with the technologies with respect to public health and the environment.

Environmental benefits. Diversions of land and water to agriculture are the greatest threats to the world's biological diversity. Worldwide, agriculture accounts for 38 percent of global land area, 66 percent of water withdrawals, and 85 percent of consumptive use of water—water that is no longer available for immediate use elsewhere. Agriculture also affects bio-diversity through water pollution and atmospheric transport—for example, by releasing excess nutrients, pesticides, silt, and greenhouse gases into the environment. Land-based agricultural activities also affect aquatic, coastal, and marine systems. In the future, the world's biggest environmental challenge will be to meet the growing demand of a larger and wealthier population for food, nutrition, fibre, timber, and other natural resource products while conserving habitat and bio-diversity. Biotechnology could be key to reconciling these opposing goals.

Worldwide, there will be a tradeoff between increased agricultural productivity and increased land conversion for the next 50 years. Assuming that the human population will be 8.9 billion in 2050 and that crop production per capita will grow at the same rate between 1997 and 2050 as it did between the early 1960s and late 1990s, and if conventional agricultural productivity increases 1 percent per year between 1997 and 2050, then 325 million hectares (800 million acres) of forests or other habitat will have to be converted to new cropland.

That's a 21.5 percent increase over the 1,510 million hectares of existing cropland. This would inevitably increase pressure on terrestrial biodiversity. Given existing, but unused, opportunities to increase productivity, such increases are plausible with conventional agriculture. However, if biotechnology increases productivity by 1.5 percent per year—corresponding to a cumulative 30 percent increase in productivity by 2050 due to biotechnology alone—then agriculture could return 98 million hectares of current cropland to nature. An increase to 2 percent per year would return 422 million hectares. Numerous GM crops, in various stages between research and commercialization, could make such gains plausible.

GM crops could be bioengineered to tolerate poor climatic and soil conditions, including acidic soils, drought, and high salinity levels, conditions which prevail in many developing countries. For instance, 43 percent of tropical soils are acidic, and one-third of the world's irrigated land has been lost to salinity. GM crops can also make agriculture more environmentally sustainable. Higher productivity leads to less cropland, less soil erosion, less silt, and less carbon in the water and the atmosphere. GM crops could also be bioengineered to fix their own nitrogen, increase nitrogen and phosphorus uptake from the soil, and resist pests. Such crops would lower environmental loading of nutrients, pesticides, and nitrous oxide.

In the United Stares, for example, in 1999, cultivation of corn genetically modified to express the Bacillus thuringiensis (Bt) gene—which repels pests without the need for chemical pesticides-reduced pesticide usage 400 thousand hectare-treatments. Production also increased by 2.3 billion liters, equivalent to the output of 200 thousand hectares. Bioengineering could also enhance feed to reduce phosphorus in animal waste, lowering its run-off into streams, lakes, and other waters. Similarly, in 1999, the use of the herbicide resistant "Roundup Ready" soybean reportedly lowered pesticide usage by 8 million hectare-treatments, while simultaneously increasing yields and facilitating no-till cultivation, a very effective method of containing erosion.

Public health. Sufficient food is the first step to a healthy society. Despite unprecedented progress in the last half century, 825 million people still suffer from hunger and malnourishment, and more than 2 billion people suffer from various micronutrient deficiencies. Worldwide, GM crops could reduce public health problems arising partly or wholly from deficiencies in the quantity or nutritional quality of food, from malnutrition to heart disease.

Cumulatively, in 1995, hunger and malnutrition were responsible for 6.6 million, or 54 percent, of the deaths worldwide in children younger than five years, stunting in 200 million children, and clinical xerophthalmia—a condition of the eye, caused primarily by a lack of vitamin A, that causes thickening and drying of the conjunctiva and cornea and could lead to blindness—in about 2.7 million people. Bioengineering can increase the quantity as well as the nutritional quality of food, as exemplified by "golden rice," which is rich in beta-carotene, a precursor to vitamin A. Crossed with another bioengineered strain, such rice could help reduce vitamin-A and iron-deficiency-related deaths and diseases globally.

Bioengineered crops can also help battle the so-called "diseases of affluence." In 1998, these diseases—heart disease, hypertension, and cancer—accounted for 4.8 million deaths in high-income countries, and 14.9 million deaths in the low-and middle-income countries. GM crops that could reduce this toll include low-fat soybeans, higher-protein rice, tomatoes with increased antioxidant content, potatoes with higher starch than conventional potatoes, and vitamin-fortified fruits and vegetables. Finally, pest-resistant GM plants can—by reducing the amount, toxicity, or persistence of pesticides—reduce accidental poisonings and other health problems affecting farm workers.

Weighing the Costs

In spite of the estimated benefits of deploying bioengineered crops, there has been much hue and cry from concerned scientists and the public about the risks inherent in this new agricultural technology. Some fear that GM crops may encourage pesticide-resistant insects and superweeds, hurt beneficial insects, and harm human health.

Perhaps the most familiar environmental concerns about GM crops centre on pest-resistant and herbicide-tolerant crops. One risk is that target pests will become resistant to toxins produced by pest resistant GM crops, such as corn or cotton. However, recent studies from Arizona, Mississippi, and Australia indicate that bollworm, for instance, did not increase its resistance to Bt toxin produced by a GM Bt cotton. Conventional strategies to deter pest resistance to conventional pesticide, such as maintaining refuges of non-Bt crops, crop rotation, and expanded monitoring, can be adapted for GM crops. Other strategies include developing crops with more than one toxin gene acting on separate molecular targets and inserting the bioengineered gene into the chloroplast—the structure in the cells of green plants that is critical for photosynthesis. These alterations allow more Bt toxin to be expressed, so that more of the pest insects die and fewer live to develop resistance. Another source of risk is that Bt plants could harm nontarget species. Studies in non-field conditions show higher mortality in Monarch butterflies fed with leaves dusted with pollen from Bt corn. Studies have also shown that the larvae of green lacewings that ate corn borers fed with Bt corn were more likely to die. However, whether—and the extent to which—these studies represent real world conditions is debatable.

The U.S. Environmental Protection Agency recently concluded that the weight of evidence indicates no hazard to wildlife from the continued registration of Bt crops, and Bt corn is unlikely to cause widespread harm to Monarchs. [18] More importantly, the inadvertent effects due to airborne dispersal of pollen or leakage into the soil through the roots of Bt crops could be virtually eliminated by bioengineering genes into the chloroplast rather than into nuclear DNA. Bioengineered herbicide or pest tolerant crops might also escape into wild relatives leading to "genetic pollution" and creating "superweeds." Since that would defeat the very justification for such GM crops, farmers and GM seed vendors have substantial economic incentives for preventing the creation of herbicide-tolerant superweeds and, failing that, keeping such weeds in check.

Gene escape is possible if GM crops are planted near wild relatives. However, the most common GM crops—soybeans and corn—have no wild relatives in the United States. Also, conventional breeding has rendered many crops—corn and wheat, for example—"ecologically incompetent," that is, the hybrid forms in common use cant easily reproduce in natural conditions and are unlikely to cross pollinate with other species. A 10-year British study of four herbicide-tolerant or pest-resistant GM crops—oilseed rape, corn, sugar beet, and potato—shows that they are no more invasive or persistent in the wild than their conventional counterparts.

This study also provides reassurance that "superweeds" perhaps can't invade natural ecosystems, and con-firms that such GM plants have no competitive advantage in natural systems. Moreover, had any herbicide-tolerant or pest-resistant weeds begun to spread, available crop management techniques—such as applying an herbicide to which the GM plant had not been engineered to b e resistant—could have been used to control them. Finally, the chances of gene escape from GM to non-GM crops can be further reduced by maintaining a buffer between the two crops.

Of course, gene escape could be limited with greater certainty if the GM plants were engineered to be sterile or prevented from germinating, using, for instance, terminator technology." An alternative approach would be to insert the gene into the chloroplast, which would preclude spread through pollen or fruit, as well as prevent root leakage. Finally, there is a concern that, in the quest to expand yields, herbicide-tolerant and pest-resistant GM plants would work too well and further simplify agricultural ecosystems, decreasing biodiversity. Some fear that the new genes inserted into GM plants could be incorporated into a consumer's genetic makeup. However, plant and animal DNA has always been a part of the daily human diet. In fact, an estimated 4 percent of human diet is composed of DNA, which is naturally present in meats and plants.

Another concern is that GM crops could trigger allergies in unsuspecting consumers. One to 3 percent of adults and 5 to 8 percent of children in the United States suffer from food allergies, and each year food allergies cause 135 fatalities and 2,500 emergency room visits. But GM foods can and have been tested, and most importantly rejected, prior to commercialization for their allergic potential. In fact, the precision of bioengineering could be used to render allergenic crops non-allergenic.

Finally, antibiotic resistant markers used to identify whether a gene has been successfully incorporated into plants could accelerate the trend towards antibiotic-resistant diseases. But this threat is slight compared to that due to overuse of antibiotics in livestock feed and by humans. Moreover, alternative markers have been devised for many crops, with more in the offing.

Applying the Principle

Clearly, there are risks associated with using or not using GM crops. The framework presented earlier offers a way to evaluate actions that could result in uncertain costs and uncertain benefits. Ideally, each criterion should be applied separately to the public health and to the environmental consequences of using or not using GM crops. However, since there are variations in the severity, certainty, and magnitude associated with the various competing costs and benefits regarding each of these sets of consequences, I will apply several criteria simultaneously.

The human population could increase 50 percent between 1998 and 2050, and with it undernourishment, malnutrition, and their consequences, assuming global food supply increases by a like amount and all else remains equal. Thus, unless food production outstrips population growth significantly over the next half century, billions of people in the developing world may suffer from undernourishment, hundreds of millions may be stunted, and millions may die from malnutrition.

Based on the sheer magnitude of people at risk of hunger and malnutrition and the degree of certainty attached to their public health consequences, limiting GM crops by limiting the rate at which food production can expand, will almost certainly increase death and disease, particularly among the world's poor. By contrast, the negative health consequences of ingesting GM foods—for example, the effects due to ingesting transgenes—are speculative. Other risks are relatively minor in magnitude, such as a potential increase in antibiotic resistance. Some are speculative and relatively minor, such as increased incidence of allergic reactions. Moreover, as noted, many of these impacts can be contained, if not eliminated.

Thus, based on the uncertainty, expectation value, and adaptation criteria applied either singly or in conjunction, GM crops must be favoured over conventional or alternative agricultural production systems. Hence, the precautionary principle requires that we continue to research, develop, and commercialize—with appropriate safeguards, of course—those GM crops that would increase food production and generally improve nutrition and health, especially in the developing world.

Some have argued that many developed countries are awash in surplus grain, and they do not need larger harvests. But reducing those surpluses would harm public health in developing countries, which currently import about 10 percent of their cereals. Without these surpluses, food supplies would be lower and food prices would be higher, as would undernourishment, malnutrition, and associated health problems. Developing countries deficits are expected to rise because of future population and economic growth and, possibly, global warming. Therefore, developed countries surpluses will continue to be critical for future food security in developing countries, just as they are today.

The above argument against GM crops also assumes that they will provide little or no public health benefits to developed countries. But GM crops are also being engineered to combat diseases of affluence, which claim 4.8 million lives annually in the developed countries. Even a small reduction in these numbers due to GM crops would translate into relative large declines in the death toll. Moreover, the health benefits of "golden rice," for instance, do not have to be confined to developing countries. Thus, even in developed countries, the potential public health benefits of GM crops ought to outweigh in magnitude and certainty the speculative health consequences of ingesting GM foods.

Another argument against developing GM foods is that hunger and malnutrition are rooted in maldistribution and unequal access to food and not in food shortages; therefore, there is no compelling need for biotechnology. This argument tacitly acknowledges that GM crops would boost quantities more rapidly than would conventional agriculture. More importantly, GM crops do not decrease access. In fact, by increasing production, they would lower prices, which would increase access, particularly for the lower rungs of society.

One factor contributing to poor food distribution is crop spoilage. But various GM crops could increase shelf life, reducing spoilage and wastage. Moreover, nutritional quality is just as important as food quantity, and GM foods can enhance the former far easier than conventional agriculture.

Bioengineered crops do not have to be the sole solution for hunger and malnutrition. It is sufficient that they contribute to the solution and are among the most efficient solutions. If the argument that there is sufficient food is truly compelling, then it should be equally valid for arguing against increases in production using conventional technologies. Perhaps, then, developing countries like India and Bangladesh should forego increasing agricultural productivity altogether, and should focus only on improving access and distribution. For obvious reasons, no one makes this argument.

Regardless of the level of demand, GM crops would increase yields per unit of land used. And higher yields mean less land for mankind, and more for the rest of nature. Moreover, reduced land demand for mankind means lower land prices and socioeconomic opportunity costs for sequestering land for conservation and preservation, which would facilitate conservation of species and biodiversity in the wild—a major goal of the Convention on Biological Diversity.

Also, by reducing amounts of cropland, fertilizers, and pesticides, GM crops would reduce soil erosion and releases of nutrients, pesticides, carbon, and nitrous oxide emissions into the environment. Compare these tangible ecological benefits of GM crops to the more speculative environmental benefits of limiting pest-resistant and herbicide-tolerant GM crops even without factoring in the environmental costs of conventional farming. Assuming GM crops actually do harm nontarget pests and weeds more than conventional farming practices, banning GM crops may increase the diversity of the flora and fauna, but this is an uncertain benefit. In addition, the possibility of gene escape to weeds and non-GM crops is also uncertain.

Hence, with respect to the environmental consequences of the use or nonuse of GM crops, one must conclude, based on the uncertainty and expectation value criteria, that the precautionary principle promotes cultivation of GM crops. This would conserve the planet's habitat, biodiversity, and carbon stores and sinks, provided due caution is exercised, particularly with respect to herbicide-tolerant and pest-resistant GM crops. It may be argued that although gene escape to the environment is highly unlikely, under the irreversibility criterion GM crops ought to be banned. However, increased habitat clearance and land conversion resulting from such a ban may be at least as irreversible, particularly if it leads to species extinctions.

Finally, the precautionary principle supports using terminator-type technology because it would minimize the possibility of gene transfer without diminishing any of the public health or environmental benefits of GM crops.

Speed with Caution

The precautionary principle has often been invoked to justify a prohibition on GM crops. However, while this justification touts the potential public health and environmental benefits of a ban on GM crops, it ignores any harm caused by delaying or foregoing benefits that GM crops would bring. Worldwide, more than 800 million people suffer from hunger and undernourishment, and more than 2 billion suffer from malnutrition each year.

As a result, millions suffer death and disease annually. But in comparison to conventional crops, GM crops, in fact, would increase the quantity and nutritional quality of food supplies. This should reduce mortality and morbidity rates worldwide. And by increasing productivity and reducing chemical inputs, GM crops would increase land available for the rest of nature and reduce global pollution by fertilizers, pesticides, and carbon. Thus, GM crops would also better protect habitat, biological diversity, water quality, and the current climate than would conventional agriculture.

Hence, contrary to conventional environmental wisdom, the precautionary principle—properly applied, with a more comprehensive consideration of the public health and environmental consequences of a ban—argues for a sustained effort to research, develop, and commercialize GM crops, provided reasonable caution is exercised during testing and commercialization of the crops.

In this context, a "reasonable" precaution is one whose public health and environmental benefits are not negated by the harm incurred due to reductions or delays in enhancing the quantity or quality of food or the health of the environment. Any such harm to public health would be disproportionately borne by the poorest and most vulnerable segments of society.

The Convention on Biological Diversity is dedicated to conserving biodiversity, preferably through conservation in the wild. But a GM crop ban is more likely to magnify threats to biodiversity, and reduce land and water available for such conservation.

Hence, such a ban would be directly counter to the CBD's very reason for existence, and that of any subsidiary agreements such as the Biosafety Protocol. Finally, slowing down commercialization of GM crops would, by reducing the future quantity and quality of food available, arguably abridge the "right to a standard of living adequate for the health and well-being...including food, clothing" and the "inalienable right to be free from hunger and malnutrition" as specified in the Universal Declaration of Human Rights and the Universal Declaration on the Eradication of Hunger and Malnutrition, respectively, thereby violating Article 10 of the Declaration on Human Rights Defenders, which states that "No one shall participate, by act or failure to act where required, in violating human rights and fundamental freedoms." While it might be a mistake to go full steam ahead on GM crops, it might be an even larger mistake to stop them in their tracks. The wisest policy would be to go as fast as possible while keeping a sharp lookout, and staying on the track to improvements in human and environmental well-being.

4

Transgenic Agriculture and Crops

Genetic engineering is an ever-expanding field in the world of science. In the last ten years, the process of genetically altering organisms has evolved from experimentation to commercial application. It is taking the place of selective breeding for isolating desired traits in different organisms. Transgenics has also led to a better understanding of gene function, allowing for better understanding of human diseases such as Alzheimer's, diabetes, and cancer. For some, such as the starving poor and those afflicted with genetically-caused diseases, transgenic organisms are a blessing. For others, such as poor farmers and the animals themselves, genetic engineering spells the end of the world.

The Federation of European Laboratory Animal Associations defines the term transgenic animal as an animal whose genome has been deliberately modified (Margawati). This definition also holds true for any other transgenic organisms, including plants and bacteria. In fact, the principle of transgenics is the introduction of an alien gene into an organism's DNA. Generally, the gene would not occur naturally in its host organism. Transgenic science was born in 1970 when experiments were conducted to produce different strains of mice; cells originating from one strain were introduced to the embryos of another strain by aggregation or injection during the blastocyst stage (C. River). The resulting offspring showed characteristics from each strain used. These first transgenic animals were created without the use of recombinant DNA, but the process was a building block for the mechanisms of genetic engineering used today (C. River). This process is actually quite similar to that of cloning, when one organism is a genetically identical to its parent.

Transgenic Organism

A transgenic organism is an organism which has been modified with genetic material from another species. The genetic modification is accomplished by inserting DNA into an embryo with the assistance of a virus, a plasmid, or a gene gun. The embryo is allowed to develop, and the mature organism will express the DNA which has been inserted into its genome. Transgenic organisms can also pass the modification on to future generations by breeding with other members of the same species. Organisms may be genetically modified for any number of reasons. Genetic modification may be performed to make organisms more vigourous, to add resistance to specific threats, or for the goal of expressing a particular trait. It can be used for everything from adding nutrition to staple crops for the benefit of people in the developing world to making fish glow so that they can be sold as novelty pets. While species cannot interbreed, as a general rule, DNA from one species can express in another. This is because DNA codes for the same thing no matter where it is, so if a researcher can determine which part of an organism's genome codes for a particular trait and the trait is compatible with the organism being modified, the DNA can be successfully inserted to cause the trait to be expressed in a different species.

Rice with Vitamin A

Rice does not contain very much vitamin A. In the poorer parts of Asia, where rice is almost the only food of the rural population, a vitamin A deficiency is common, leading to early blindness. Now Drs. Ingo Potrykus and Peter Beyer, two genetic engineers, have transferred the genes for vitamin A from other species into rice, creating a strain of rice which is rich in vitamin A — the amount of rice in a typical third world diet could provide about fifteen percent of the recommended daily allowance of vitamin A, sufficient to prevent blindness. Now that a few plants with this trait have been created, they are being cross bred with other varieties of rice using conventional breeding techniques, as has been done for centuries.

Such cross breeding could further increase the vitamin A content. The development of rice with vitamin A was carried out at the Swiss Federal Institute of Technology, making free use of patented technology and of the earlier research which had established the basic facts about how plants synthesize vitamins. The corporations holding the various patents all agreed to cost-free use of their patents as long as the rice was to be provided free to poor third-world farmers. The new rice strain was then turned over to the International Rice Research Institute, a non-profit organization based in the Philippines, where it will be evaluated for its adaptability to various growing conditions, food safety, and environmental impacts, etc.

The IRRI preserves thousands of varieties of rice with different genetic characteristics, so the new strain can be cross bred to produce varieties suitable for almost any locality. The result is that rural Asians can soon expect to retain normal eyesight. Genetic engineers also intend to produce a rice variety rich in iron, because iron-deficiency anemia is a common problem in the same rural populations. But this is a more difficult problem than increasing rice's vitamin A content. Rice contains a substance called phytate. Phytate prevents the body from absorbing iron, so it does little good to breed for increased iron content, and the rice plant cannot reproduce without adequate phytate in the grains. Dr. Potrykus hopes to be able to find a gene coding for a protein that will break down phytate when the rice is cooked.

No-Till Agriculture

The world's biggest environmental problem is loss of topsoil to wind and drainage. The US experienced its dust bowl during the early part of the 20th century as a result of ploughing up the prairie. The problem is much worse in tropical soils, which may have a thin, inches thin, layer of topsoil above a type of soil which, when ploughed, turns into a non-porous concrete-like substance. One field, one crop, once. The problem, both in the prairie and in the tropics, is deep ploughing, which kills weeds which would otherwise crowd out the desired crop. The solution is called low-till agriculture.

The soil is broken up but not deeply ploughed. Weeds are killed instead by herbicides. A herbicide of choice should be cheap, quickly biodegradable and non-toxic. An excellent choice is a chemical called glyphosate, except that glyphosate kills the crops as well as the weeds. So genetic engineers found a gene which lets plants tolerate glyphosate, and transferred it into soybeans.

Today, 63 per cent of the soybeans grown in the US are glyphosate tolerant, allowing soil saving no-till agriculture on half the US soybean acreage. There is another advantage to no-till agriculture. There are lots of plant residues beneath the ground, both root systems and humus transported by earthworms. Ploughing brings this material to the surface, where it can oxidize.

Carbon dioxide is created, a greenhouse gas. So transgenic soybeans are a positive factor in postponing global warming. Approximately four tons of carbon dioxide are retained in the soil per acre per year. This saving is applicable to the accounting of CO_2 reductions in the Kyoto Treaty on Climate. In fact, any technology which reduces the need to plough, spray, or till crops will reduce carbon dioxide emission. Consider a tractor pulling a ten foot wide harrow over a square mile of agricultural land. Simple arithmetic shows that the tractor will travel 528 miles, all the while burning gasoline. Perhaps you are thinking that even if ploughing has disadvantages, herbicides don't sound very good either. The very word means "plant killer". But that is not the choice. Traditional soybeans are also grown using herbicides, most of which are far more toxic than glyphosate. On the average, the transgenic soybeans actually use 30 per cent less total herbicide than conventional soybeans. So environmentally, this is a no-brainer.

Witchweed Control

Farmers in east Africa are plagued by a devastating parasitic weed called Striga, or witchweed. Farmers are used to dealing with weeds that grow in the soil alongside the crop and compete for nutrients. From time immemorial, they have dealt with those weeds by pulling them up by hand. Less neighbour intensive methods like spraying and ploughing are now common. But none of these methods work for the witchweed. Striga attacks plants directly, underground, even before the weed has emerged above the soil surface. It sucks nutrients from the seeds and the roots of the crop. In some parts of Africa, the striga parasite destroys as much as 80 per cent of the crop yield.

But now that a herbicide resistance trait can be transferred to a crop, scientists in Israel and Kenya, working together, have demonstrated a new strategy for striga control. Before planting the crop, they soak its seeds in a herbicide. The seeds are unharmed, but they become poisonous to the striga parasite. The seed germinates and sprouts without interference. By the time the crop is harvested, the herbicide has decomposed and disappeared. Their demonstration used herbicide resistant transgenic corn. The same strategy would probably work with Africa's other important grains, sorghum and millet. Soaking seeds would use far less herbicide than spraying it on the ground, and the complex spraying apparatus would not be needed. This is a significant consideration in Africa, where so many farmers are too poor to own expensive equipment.

Saving the Banana

Wild bananas have seeds. They reproduce sexually, like beans and oak trees. But they aren't easy to eat. Bananas grown on plantations have no seeds. They are reproduced by taking cuttings from older banana plants. Cultivated bananas are seedless because they have three of each type of chromosome instead of the normal two of each type. Such plants are called "triploid". They are always sterile. Genetic triploid freaks arise from time to time in nature, but modern breeders can also use chemicals or electric shocks to create triploid mutant cells. Bananas have been cultivated for many thousands of years and there are about three hundred different banana varieties. Each variety was developed by crossbreeding wild bananas. Whenever a promising variety was been produced, the breeder caused it to be triploid, hence seedless.

That plant was cloned by propagating cuttings, and it became the parent of its variety. All bananas plants of the same variety are genetically identical, like identical twins. Of the three hundred varieties, only one single variety completely dominates international trade. It is called

Cavendish. It is possible that you have never seen a banana other than a Cavendish banana. Certain kinds of fungus can infect and kill banana plants. In many parts of the world, Cavendish banana plants are being attacked by a fungus called black Sigatoka. Since wild bananas can reproduce sexually, they are not all identical and some wild bananas can resist black Sigatoka. If the black Sigatoka fungus is present in a region, the resistant banana types become prevalent. But the Cavendish banana plants have no resistance.

They are all identical so they all die. To grow bananas commercially, growers must spray their plants with fungicides. Year after year, the black Sigatoka fungus has been evolving resistance to these fungicides, so growers have to spray more and more fungicide each year. Approximately one third of the cost of raising a banana is the cost of spraying it with fungicides, and it gets more and more costly each year. A form of black Sigatoka banana disease, now spreading around the world, can tolerate all known fungicides. Soon it will attack bananas in Central America and the Carribean islands, the heartland of banana culture. The Cavendish banana will become virtually extinct. Agronomists estimate that this will happen within ten years.

This is not a fairy tale. It has happened before. Forty years ago, the most popular variety of banana was one called Gros Michel. But Gros Michel was susceptible to a fungus called "race 1 Panama disease". Now it is gone. Cavendish bananas, which are not susceptible to race 1 Panama disease, replaced them. (A related fungus, race 4 Panama disease, can infect Cavendish banana plants. At present it is only found in Malaysia and some nearby countries.) In turn, some other variety of banana could replace the Cavendish. It would look different and taste different but it would still be a banana.

There is only one practical way to save the Cavendish banana. It must be given some combination of genes from wild bananas which are not susceptible to the fungus. But this can't be done by any natural technique. Cross-breeding can create other banana varieties because wild bananas reproduce sexually. But Cavendish bananas reproduce only by cloning. Conventional breeding would have to rely on rare mutated Cavendish banana plants which can produce seeds and which can therefore be crossbred, in theory.

Even the mutated plants produce only a tiny number of viable seeds, as few as two or three seeds in a hundred pounds of bananas. No banana breeding experts think that they can breed fungus resistance into a Cavendish banana variety in only ten years. Biotechnology can rescue the Cavendish banana. Scientists in Belgium have used genetic engineering to transfer some fungal resistance genes from wild bananas. The transformed Cavendish plants are not susceptible to the fungus and they can then be reproduced into nursery stock by the usual method of taking cuttings.

The Eggplant in Winter

The edible part of an eggplant is formed from the ovary of its flower. In this way, it is like the edible flesh of an apple, a pepper or grape. When we eat these fruits, we discard the seeds. But the plants only transform their ovaries into fruits when they start to produce seeds, although in the case of an eggplant, its seeds are so tiny that we ignore them. Eggplants will only set seeds in warm weather, so to grow them in the winter in an unheated greenhouse, the grower must use a chemical to trick the plant into beginning fruit development without setting seed. Such fruits do not grow very large or very fast under these conditions. So eggplants are expensive in the winter.

But now scientists in Italy have transferred two genes into a variety of eggplant, which not only allows the plant to set fruit in cool greenhouse conditions without chemicals, but also increases productivity of the same plant in either hot or cold weather. The eggplant variety that the Italian scientists created is seedless. One of the two transferred genes is a switching gene which is turned on only in the ovary part of a flower. That gene turns on the other transferred gene, which makes a protein involved in synthesizing a growth hormone. The growth hormone makes the ovary grow into the fruit, just as it would have done in a traditional eggplant making seeds. Neither gene requires either seed setting or warm weather. Where does one get seeds to produce large numbers of seedless eggplants? The transformed plants produce pollen, so they can be crossed with traditional eggplant varieties and the hybrid produced by that crossing has the seedless and self-starting property. The scientists report productivity increases of 37 per cent for the new variety, and they believe that the seedless type would be more marketable.

Frustrate Transgenic Agriculture System

Although there are legitimate reasons to oppose genetic engineered agriculture, or at least to demand the most careful controls, there are a community of opponents who have taken their opposition beyond what is ethical. I am not talking about a reasonable opposition expressing the concerns summarized earlier, but rather about an opposition for which the ends justify the means, including lies, vandalism, etc. I need to stress this point. There are many people, sincerely opposed to genetic engineered crops, whose ethics I do not question. I believe that most of the opponents of biotechnology would fall into that category. Yet, in too many cases, those sincere concerns are based entirely on misinformation which originated in deliberate lies and fear mongering.

We are here exposing the ethics of the people who have created the lies. These opponents must have some motive, and it seems that an alliance has emerged between at least four groups, each with its own agenda. First there are people who sincerely believe that genetic engineering is ethically wrong, and that anything they do to stop it from happening is therefore right. Second there are foreign governments and their constituencies who are worried about American domination of agriculture. Closely related to this are advocates of organic agriculture who seem to be engendering public fear to make their own products more salable.

Third, there are environmental groups who have been misled by a radical fringe and have become willing to do anything to stop genetic engineering agriculture. The most conspicuous among these groups is Greenpeace. Finally, there are people opposed to capitalism or to large businesses dominating agriculture. Opposition from environmental groups is particularly frustrating to me. Most measures which benefit the environment require people to give up something, and they don't like to do that. Recycling is nearly painless and saves money, but many people won't make the effort. A few degrees adjustment of a thermostat could save vast amounts of energy, but most people would rather be comfortable. We end up settling for half measures. But here is a technology that benefits the environment without asking people to give up anything, and its biggest opposition comes from environmental groups.

Misinformation about Food Safety

There is a deliberate campaign to frighten people about the safety of the food supply. This campaign has worked successfully in England and in much of Europe. Dr. Arpad Pusztai, who worked at the Rowett Institute in Aberdeen, Scotland, performed an experiment. It began when a gene was transferred from a poisonous plant, the snowdrop, into a potato. The transferred gene specifies the production of a poisonous compound called lectin. Dr. Pusztai proceeded to experiment with rats. Some rats were fed with the raw potatoes which were genetically engineered to contain the poison. The rats in the control group were fed ordinary raw potatoes and were also given the amount of lectin poison which the first group of rats would have gotten from eating the transgenic potatoes. Both groups of rats developed malformed organs, and there was no statistically significant difference between the rats who consumed the poisonous potatoes and those who consumed the poison.

However, Dr. Pusztai claimed that his data showed that the rats who ate the genetically modified potatoes had more deformed organs. No scientific journal would publish Dr. Pusztai's interpretation, and his institution would not support him. He hired an independent statistician to review his data, who also considered the data to show no difference between the two groups. But Dr. Puzstai held to his opinion. Eventually the disagreement became serious enough that his connection with the Rowett Institute was ended. The opponents of genetic engineering, mostly in England, have blown this result into a cause celebre. Dr. Pusztai is portrayed as muzzled by the scientific establishment, although the British medical journal Lancet eventually published Pusztai's paper over the recommendations of its reviewers because of the widespread public interest.

The British tabloid press covers this story continuously, with lurid photographs of deformed rat organs. The potatoes genetically engineered to be poisonous became synonymous with all transgenic food, called Frankenstein food in the tabloids. There are varieties of potatoes, bred to be eaten, engineered to resist insects, viruses and fungi. All these varieties have been fed to rats and have never harmed them. The opponents of transgenic food have no explanation for that - they are content to use one probably misinterpreted experiment with potatoes nobody will ever eat, to stir up doubts about food safety. During safety testing of transgenic tomatoes, Dr. Belinda Martineau discovered that when rats are fed huge amounts of tomato paste they can develop stomach lesions.

It doesn't matter whether the tomatoes are transgenic or not. But before there were transgenic tomatoes, nobody had ever fed rats such large doses. Martineau's results were reported to the FDA but opponents routinely call this a "cover-up" of a health hazard with transgenic tomatoes. As another example, I read an op-ed article in the Boston Globe in August of 1999, written by Paul Billings, a member of the board of the Council for Responsible Genetics. The gist of the article is that dangerous untested foods are being foisted upon an unsuspecting American public, by mad scientists. Every genetically engineered crop has been tested for safety. The testing has been much more extensive than that for any other foods, including foods developed by radiation induced mutation. Dr. Billings knows this. He knows about the government testing rules that establish the safety of each individual crop.

He wants genetically modified food to be tested as strictly as drugs are tested. Billings' op-ed article contains only one 'fact' that most people would not have known before - the rest is either his opinion or just wrong. He says that transgenic soybeans have been shown to be deficient in a certain unidentified nutrient. It is not easy to track down the source of this 'fact', but I did it. The nutrient in question is a phyto-estrogen (also known as a phytosterol). Although phyto-estrogens are not essential to human health, there is some indication that they help prevent cancer. The study that indicates that transgenic soybeans are deficient in phyto-estrogens comes from Dr. Marc Lappe, who wrote the book "Against The Grain", a polemic against genetic engineering and especially against the Monsanto Co., the leading company in the field, which developed the soybeans in question. Here is how Dr. Lappe established that genetic engineered soybeans are defective. Understand that there are dozens of varieties of soybeans with the herbicide tolerance trait and over a hundred varieties of conventional soybeans. Dr. Lappe compared one conventional variety with one transgenic variety. He found a 12 per cent difference. But individual soybean varieties vary by more than 100 per cent in their phyto-estrogen content.

Also phyto-estrogen content is not stable. It declines with storage, by much more than the 12 per cent difference. But Lappe at least reports his data along with his biased interpretation of it. Dr. Billings reports only the interpretation without any indication that it comes from a biased scientist whose own data show insignificant variations in a nutrient whose role in human health is not even firmly established. Dr. Billings seeks only to mislead people. Not one reader in a thousand would do what I did, track down the data.

His purpose is to plant a little seed of doubt about food safety, hoping that it will fester in our minds, mingle with similar misinformation, and eventually become accepted fact. Not everything misleading is false. Propagandists are very skilled at making true statements that seem to imply something quite different.

For example, they frequently say that the FDA's rules about testing genetically modified food are voluntary, implying that some testing doesn't get done. Under current law, FDA has no authority to require safety tests for any food, transgenic or conventional, although it can prevent the sale of foods it considers unsafe. But, in fact, each developer of a transgenic crop has consulted with FDA and performed every test FDA suggested. They also like to say that safety test results are trade secrets. True, it would be legal to keep test results secret.

But no developer has done it. Earlier I mentioned a problem with a soybean with a Brazil nut gene. People allergic to Brazil nuts should not expect to have to avoid soybeans, but the allergen was identified by testing and therefore the modified soybeans were never created.

Also, the scientists who demonstrated this published their results in a scientific journal, and this led the FDA to not allow any gene to be transferred into a food from a species known to cause allergies. That should be seen as evidence that the genetic engineers are responsible people, and that testing is working well. But the unethical opponents of biotechnology routinely present this episode, carefully worded, as if there was a near disaster, revealing gross problems with the current regulatory system.

More From The Sun

Plants derive energy from sunlight and use it to make sugar from carbon dioxide and water. This is photosynthesis. Scientists still do not have a complete understanding of how photosynthesis happens, although they know most of the steps. They know many of the genes which create the proteins needed for photosynthesis. They also know that there are differences in photosynthesis from one species to another. It happens that corn is an overachiever.

Corn plants make more sugar per unit of sunlight than any of the other grains. An international team of scientists from Japan and from Washington State has transferred three of corn's photosynthesis genes into a rice plant. Early indications are that the transformed rice is more productive than the original rice variety. A more important potential application may be the development of very fast growing trees. If global warming cannot be prevented by adding less carbon dioxide to the atmosphere, by burning less coal and oil, the only alternative is to depend on processes that remove it. Number one on that list is growing new trees.

Anything that makes agriculture more efficient can make more land available for growing trees. Anything that makes those trees grow faster removes more carbon dioxide from the atmosphere. For many environmentalists, preventing global warming is the highest priority. But proposed measures to restrict burning fossil fuels have encountered fierce political resistance. Opponents claim that such restrictions would be cost too much money, and would cost people their jobs, their comfort and their prosperity. By contrast, nobody loses anything if carbon dioxide is removed from the atmosphere by growing new trees.

Toxic Soils

Some soils are poor for plant growth because their mineral content is toxic. A high aluminum content is the most frequent problem, especially in acidic soils. But it has been possible to identify a few genes which enable some plants to extract aluminum compounds from soil and sequester them harmlessly in their fibrous parts. Recently, Florida scientists discovered a type of fern which can extract arsenic from the soil, although they do not yet know how the fern does this. But other teams have identified genes that can enable plants to remove cadmium, zinc and mercury from soils. By transferring such genes to fast growing plants, it should be possible to clean up some toxic soils in much the same way as we can use bacteria to clean up oil spills. In the nearer term, there is the work of Mexican scientist

Luis Herrera Estrella. He transferred into corn a gene that allows the plant to overproduce a natural chemical, citric acid, which it then excretes through the roots. Citric acid binds to aluminum and prevents the plant from taking it up from the soil. Herrera's approach is not to extract aluminum from the soil but to prevent it from passing from the soil to the plant. A much larger problem is salt-contaminated soil caused by irrigation. Rainwater is very pure, but water borrowed from rivers contains some dissolved salt. Over many years of irrigation, the salt accumulates. But water cannot get from soil to roots if the soil water is saltier than the intracellular water.

In fact, water goes the other way, from plant to soil, and the plant dies. A gene was identified in a relative of cabbage. This gene enables the plant to pump salt from the soil into an isolated part of a cell, called a vacuole, where it is stored without harm to the plant. When salt is thus removed from the soil around the roots the plant can then take up the less salty water. The salt-tolerance gene was experimentally transferred to a tomato plant, where a control gene keeps it turned on all the time. The resulting tomato plant is able to grow well in salty soils. Happily, the fruit is not high in salt, but the plant's stems, leaves and roots are loaded with salt, so after the growing season the plant parts could be shipped elsewhere, making the soil become less salty each year. It's one more case of an environmental problem that can be solved by gene transfer.

Biological Pest Controls

Farmers, for very obvious reasons, would prefer not to use pesticides. They cost money and they are dangerous to use. Farmers much prefer Integrated Pest Management (IPM), a system that combines many different methods of suppressing crop pests, including encouraging predatory insects. Farmers even buy them. Agricultural distributors can supply such insects as ladybugs, praying mantids, lacewings and parasitic wasps. Integrated pest management includes using pesticides when other means are insufficient. But when a crop is sprayed with conventional insecticide, the harmful insects are not the only victims. Predatory insects may also be wiped out. Without any predators available, the pest populations can recover quickly, so that a second application of pesticide is required, which also kills the insect predators. This vicious cycle could be broken if genetical engineers can develop predatory insects resistant to common pesticides.

Rust Resistance

To a plant scientist, rust has nothing to do with oxidized metal. It is a plant disease caused by a fungus. It blights all the cereal crops, barley, wheat, oats, corn, millet and sorghum, but not rice. The rust fungus reproduces itself by forming club shaped cells called basidia. Each basidium bears four spores. When the spores are ready, they are released and carried by the wind. The fungus infecting a single grain of wheat can easily produce millions of spores. The spores are so light that they can travel several times around the world before falling to the ground. Although some varieties are more resistant to rust than others, no variety is immune. But rice must contain some combination of genes that confers immunity to rust. If these genes can be identified and if their function can be deciphered, it should be possible to transfer them to other cereal crops and end, once and for all, this most important cause of famine.

Fast Growing Trees

Making paper requires large amounts of natural cellulose. Some can be derived from recycling, but most of our paper is made from freshly cut trees. The best trees for paper-making are fast growing softwoods with low resin content, like aspens. Genetic engineers have transferred genes for pest resistance and herbicide resistance into aspen and have tinkered with the genetic switches that promote growth to create a fast growing aspen that could supply our paper needs using considerably less land. The paper-making process must bleach out the brown colour of lignin, one of the components of wood. The bleaches used to be dumped in the nearest rivers, an important and highly visible kind of pollution. This is no longer allowed, but the disposal of chemicals from paper mills is still a major headache. At the Michigan Tech University, researchers have reduced the lignin content of aspen so that fewer chemicals are needed in the paper making process.

Fast Growing Fish

Most of the salmon we eat are caught wild, but some are grown in farm ponds. It takes about three years for a salmon to grow from fingerling size to optimum marketing size. In wild salmon a control gene turns on the gene for growth hormone, but only in the pituitary gland and primarily in warm water. So genetic engineers used a different control gene to turn on a growth hormone gene in cold water. That control gene was transferred from an ocean pout, and it originally turned on a gene for a protein that helped the pout tolerate very cold water. The resulting creature looks and tastes just like the wild type salmon but it grows three times faster so it ought to be cheaper to produce. Wild salmon are now under environmental pressure from overfishing and because many of the streams where they lay their eggs are either polluted or inaccessible. If farmed salmon can economically replace more wild salmon, the pressure on this desirable species could be reduced dramatically. There's a need for fast growing fish in rice growing regions. Rice is planted in standing water, but it is harvested from dry ground. With a slow growing rice farmers often raised fish in the rice paddy alongside the young plants. But newer strains of rice mature almost twice as fast as traditional varieties. Although this lets farmers grow more crops per year, unfortunately the rice paddies are not flooded long enough to raise fish. If genetic engineers were able to make fish grow faster, the farmers could again exploit this valuable protein resource.

Consumer Traits

Most of the traits in the examples mentioned so far have been targeted at the producer. The no-till soybeans are cheaper to grow because ploughing costs money. The cotton is cheaper to grow because chemicals cost money. The chymosin from yeast is cheaper than chymosin from calves. The consumer doesn't know how much water or fertilizer the farmer used, or whether a salmon is one year old or three. The vitamin enriched rice is the only one of the examples where the final product is better, rather than cheaper, for the person who eats it. But in the future we can expect to see "consumer traits". One of the first to appear will be potatoes genetically engineered to have a higher percentage of solids.

If you love french fried potatoes but don't like the calories, you will love the new potatoes. They will absorb less oil but stay crispier longer. Another valuable trait coming soon is coffee beans without caffeine. Caffeine is now removed from coffee by a chemical treatment, invented by German chemist Ludwig Roselius. Decaf is the only coffee I drink so I am looking forward to cheaper decaf coffee. In addition, many common foods are not safe for everyone.

For example, peanuts cause a life threatening allergy in some people, especially children. Allergens are unusual proteins which are digested very slowly. Someone who has an allergy to a widely used food needs to read the small print on product labels, quiz the waiter in a restaurant, etc. If a child has the allergy, the problem is that much harder to manage. But peanuts or other crops are now being genetically engineered to eliminate these allergens. USDA scientists have identified the principal allergen in soybeans and have successfully developed modified soybeans which do not produce that allergen. It is much easier to deactivate a gene, once its function is discovered, than it is to transfer a gene from one organism to another. The same antisense trick that was used to delay ripening can be used to suppress synthesis of an allergen. Therefore, once a gene has been identified which codes for an allergenic protein, the technology to eliminate that allergen from food crops is relatively easy, unless the allergenic protein is important to the life processes of the plant.

Lawns Don't Need Frequent Mowing

In addition to food and fibre, genetic engineers are working to modify grass. Farmland constitutes mankind's biggest footprint on the earth, but lawns, athletic fields, golf courses, etc. have the biggest footprint in some communities. Environmentalists have many criticisms of lawns. They must be mowed regularly, which uses gasoline, creates noise pollution, and takes up people's time. In addition, the growing grass uses a very significant amount of water, fertilizers and pesticides, to make longer leaves which are then cut off by the mower. But genetic engineers are trying to develop a variety of grass which reaches a desired length and then dramatically slows its growth. Combined with pest resistance genes, this new grass would be almost as simple to maintain as astroturf.

Transgenic Crops

The application of recombinant DNA technology has revolutionized plant breeding in recent years. Genetic transformation in plant breeding greatly increases the gene pool and broadens the scope of genetic changes that modern breeders may draw upon. The first generation of genetically modified (GM) crop plants hold great promise in carrying new or improved agronomical traits that require less intensive farming methods or which have improved quality traits. These GM crop plants have been produced to improve farming production systems, aiming, for instance, to reduce negative impacts on the environment, such as fertilizers, herbicides and pesticides.

In general they are of little direct interest to the consumer. However, the genetic manipulation of the primary and secondary metabolism of plants offers distinct possibilities in the development of nutritionally improved products, raw materials with added pharmaceutical value, and crops with real health benefits for the consumer. These are second-generation GM plants. Recently, several prototypes in molecular plant breeding with novel traits for (non-) food and industrial applications have been introduced into the market, and many more GM crops have the prospect of commercial exploitation in the near future.

The safety and wholesomeness of these GM crop plants raises questions beyond those posed by conventional foods. In this context, traditional plant breeding of most food crops has conformed to the standard 'generally recognized as safe' (GRAS). In some instances safety is assumed to be assured during performance and quality testing by breeders (*e.g.* glucosinolates and erucic acid in rape seed). However, the process is largely focused on the comparison of agronomic and phenotypic characteristics, such as colour, maturity, yield or disease resistance. Products with an unusual taste or that have harmful effects following ingestion or skin contact have normally been rejected from the breeding programme. However, the safety of conventional plant varieties has usually been presumed from the assumption that prudent consumers will avoid those species once it is known that they cannot consume them without adverse health effects. When the commercial production of GM plants started, it sparked a debate centred on issues associated with concerns on environmental, human and animal safety matters. Many people questioned whether the 'simple' GRAS approach, based on historical and empirical experiences, is sufficient to guarantee food safety of the GM crop plants.

It is also asked whether the risks of chronic exposure of humans and animals to GM crops are adequately covered, and whether large-scale breeding of GM plants will have detrimental effects on ecosystems.

Food safety issues

Although the variety registration of existing crop plants has not resulted in adverse effects in humans, the conventional assessments carried out by plant breeders was not considered to be sufficient to ensure food safety of GM crop plants. As a result, Europe issued a regulation on novel foods and novel food ingredients that sets the legal framework for the market introduction of genetically modified organisms (GMOs). This came into force on 16 May 1997.

The Novel Foods Regulation establishes a system of mandatory pre-market approvals of novel foodstuffs, including GM plants. The accompanying EU guideline 97/618/EC gave a clear indication of the types of data that are needed to form the basis of an assessment. In particular, four central safety issues have been set forth related to food and feed safety:

- The nutritional and toxicological consequences of inserted gene products;
- The potential of pleiotropic (unintended) effects in the host organism due to the insertion event;
- The allergenicity of expressed proteins and novel foodstuffs;
- The potential of gene transfer to human and animal gut flora.

This means that in transgenic insect-resistant Bt tomatoes encoding the C-terminal truncated Bt2 gene derived from a *Bacillus thuringiensis* (Bt) strain, *IAb5*, the novel protein has to be evaluated on its own merit, while the safety and wholesomeness of the 'remaining' transgenic tomato is assessed separately. Whereas, for example, in antisense RNA exogalactanase tomatoes, which have improved rheologic characteristics due to downregulation of the endogenous exogalactanase activity, the safety assessment is primarily focused on the 'remaining' novel fruit. In general, there is basic agreement on the safety issues to be addressed. One particular area of concern is the possibility of unexpected or unintended metabolic perturbations due to genetic modification that may alter, for instance, levels of nutrients and health-influencing components. The use of rDNA techniques does not necessarily result in fundamental changes in crop plants compared to the food produced by conventional breeding methods. However, it should be emphasized that a uniform international agreement for the evaluation of GMOs is urgently needed.

For example, an agreement on how to establish in practical terms the similarities or differences between the 'remaining' GM crop plant and the traditionally bred plant seems to be much more difficult to achieve. This chapter is devoted to the description of advanced strategies for the evaluation of potential unintended effects in GM crop plants. Pleiotropic (unintended) alterations in agronomic traits or composition may arise from insertion mutagenesis or as a result of metabolic effects of the novel gene product(s). The challenge is to gain the ability to discriminate between metabolic alterations due to somaclonal variations, unintended effects or natural diversity. Thereto, emphasis is given to a platform of technologies designed to identify changes in the molecular machinery of crop plants. As is the case for detection methods for GM crops and derived products, there are needs and criteria for sampling strategies, statistical models and manufacturing standard reference matehers recently.

Unintended effects

Strategies for assessing the food safety and wholesomeness of novel foodstuffs are currently in the exploration phase. Above all there is a need for uniform and harmonized quantitative methods for the testing of potential unintended effects in whole foods, such as GM crop plants are. The safety of, for example, synthetic chemicals and food additives is established by assessing separate elements of the compound in question.

Although very successful in case of single chemicals, the evaluation of a whole food product may not be a simple addition of individual assessments of the constituents. Traditional testing protocols involving labouratory animal 90-day feeding trials with (whole) food products are far from ideal. This type of animal feeding study with complex food matrices is complicated by the likelihood of nutritional imbalances leading to dietary problems, confounding factors, an insensitivity for specific endpoints, and the impossibility of using large safety margins, if any.

Therefore, in a global context several regulatory, research and industry-based bodies have proposed alternative concepts for the safety evaluation of genetically engineered foods and food ingredients. In particular the Organisation for Economic Co-operation and Development

(OECD), the Food and Agriculture Organization (FAO) and the World Health Organization (WHO) have developed concepts and principles for the safety evaluation of GMOs and derived products. Their reports concluded that risk assessments should be directed at demonstrating that a GM crop plant or derived food product is as safe as a traditional or previously authorized product. Accordingly the EU Competent Authority incorporated these general principles into their Novel Foods Regulation and accompanying guidelines. In this context, novel foods and food ingredients that are considered to be 'no longer equivalent' to traditionally bred plants are subject to labelling as defined in the Council Regulation 1139/98.

Substantial Equivalence

The term 'substantial equivalence' appears in the EU Novel Foods Regulation as a major principle in the risk assessment. It emphasizes that, within the context of the Regulation the evaluation of the safety and nutritional value of a GM plant should be focused on a comparative analysis with conventionally bred products.

OECD's Group of National Experts on Safety in Biotechnology enunciated this comparative approach towards the safety assessment of transgenic food plants as the concept of substantial equivalence. In their report, it is assumed that existing food organisms possess a long history of safe use (*i.e.* GRAS). Consequently, the most practical approach to the determination of safety is to consider whether the GM plants are substantially equivalent to analogous food product(s), if such exist. Once substantial equivalence of a novel foodstuff has been established, it provides assurance of safety that is equal to or better than that of its comparators.

Further clarification of the concept of substantial equivalence originates from a WHO sponsored workshop in 1994. It has been concluded that the establishment of substantial equivalence is not a safety assessment *per se*, but a dynamic, analytical exercise. Thus, it may mean that analysis of, for instance, gene expression patterns, global changes in protein expression and/or differences in metabolic capabilities (the metabolome) of the novel product in comparison with conventionally bred products should be performed in order to examine equivalence.

This strategy has worked satisfactorily for the assessment of the first-generation crop plants for which sufficient background knowledge is available. However, relatively less experience has been gained with the safety and wholesomeness evaluation of novel food plants for which no substantial equivalence can be established. Even in the case of single gene modification, potential alterations in metabolic pathways are difficult to identify and, consequently, the implications of the genetic modification process on the metabolism of plants are poorly understood. Moreover, data on the mechanisms by which plants regulate, for instance, their response to environmental stress factors and other conditions are remarkably scanty.

Critical Nutrients and key Toxicants

Application of the concept of substantial equivalence appears in practice to lead to various interpretations. Different requirements for risk assessment with regards to the potential of unintended effects exist as a result of genetic modification. Actual approaches entail a consideration of the characteristics of host and donor organism, phenotypic properties and toxicological evaluation and include comparative determinations of any changes in critical (anti-)nutrients and key toxicants for the food source in question. In general, the analysis of an expanded spectrum of components is unnecessary, but should be considered if there is an indication from other traits that there may be an unintended secondary effect of genetic modification.

However, compositional analyses based on single parameters using a list of crop-specific critical (anti-)nutrients and key toxicants as a minimum requirement of screening for possible variations in the content has its limitations. Limited data are typically available for the less important food components, as in less well-known crops most critical (anti-)nutrients and natural toxicants will be unknown.

Moreover, the toxicants to be assessed may be partly influenced by the cellular function of the product encoded by the inserted gene. Furthermore, the natural variation of critical comparators may mask the evidence of secondary effects due to genetic modification. Thus, there is no criterion that determines whether a difference between a GM plant and its natural counterpart is outside the literature values for the ranges of parameters that express natural diversity.

Relevant information on indicators of unexpected effects may not have been observed during the development of an edible transgenic crop species. In the future other compounds could become of importance, such as anti-oxidative, oestrogenic and anti-carcinogenic compounds. In determining critical nutrients and/or key toxicants, differences in consumption patterns and practices of processing and consumption in various geographical settings and cultures must also be recognized. It is our opinion that conclusions about relative safety and wholesomeness based on a list of selected single components may not be equally valid in a global context and in all times. Therefore, there is an urgent need to set up generic approaches using platform technologies to evaluate undesirable metabolic perturbations in GM food crops.

A Post-genome Challenge

In our view the identification of unintended effects encompasses the ability to interpret and use innovations in molecular genetics research such as genomics, proteomics and metabolomics on a crop-by-crop basis. This will establish a whole data package directed towards a holistic view on possible unintended side effects due to genetic modification. In essence, the focus is shifting towards molecular characterization to understand functional activity. Vital to this approach will be informative profiles of molecules at different integration levels, *e.g.* mRNAs, proteins including post-translational modifications and molecules at the level of primary and secondary plant products and metabolites that might be useful benchmarks for the detection of unintended effects. The relative importance of these various 'profiling' techniques in establishing data sets for assessing unintended effects is not indicated by their order and will vary from species to species. Alterations in, for example, expression levels, post-transitional modification, interactions or chemical composition do not *per se* imply that the product is less safe or even unsafe. Such a platform of technologies may determine whether unintended effects exist between the GM plant and its control(s) or whether there is substantial equivalence apart from certain well-defined expected differences.

Measures of the relative activities of various molecular constituents associated with metabolic capabilities in non- and modified plant cells under different conditions will provide new insights into how metabolism and genetic modification are orchestrated. But it should be emphasized that the fields of plant genomics and proteomics are still in their infancy.

cDNA microarrays

Over the past few years, the DNA microarray technology has emerged as a powerful high-throughput method for the analysis of gene expression. At present, microarrays for gene expression studies generally consist of cDNAs, which are physically deposited onto small glass surfaces. The major advantage of DNA microarray technology over conventional gene-profiling techniques is that it allows small-scale analysis of expression of a large number of genes in a sensitive and quantitative manner. Furthermore, it allows comparison of gene expression profiles under a large number of different conditions.

The cDNA microarray technology has already been successfully introduced in disciplines ranging from cell and developmental biology to drug development and pharmacogenomics. In the following a number of technical aspects of the technology as well as its potential value for the safety assessment of genetic modification of food plants will be discussed.

The cDNA microarrays are produced by deposition onto a solid surface of purified PCR products corresponding to specific genes.

Typically a few nanolitres of DNA solution (100-500 μg/ml) are spotted using a micro-dispensing robot. Essential information on robot construction and protocols can be found on the Internet. In general, with respect to the spotting devices, a higher density of spots is accomplished when using a passive dispenser (> 2500 DNA spots/cm^2). Thereto, glass slides covered with a positively charged layer (*e.g.* poly-l-lysine) or carrying reactive groups are most commonly used.

To analyse the expression of genes in a plant cell or tissue sample, polyA+ RNA (typically 0.5-2.5 μg is required for each reaction) is purified and labelled by reverse transcription using an oligo-dT primer incorporating a fluorescently labelled nucleotide. The labelled cDNA pool is then hybridized to the microarray. The required hybridization conditions (*i.e.* sample concentration, stringency of hybridization) and parameters, such as detection level and the correlation between transcript concentration and hybridization have been determined for each experimental setting.

The intensity of a hybridization signal, which is a measure of the relative expression level of each gene, can be read by charge-coupled device (CCD) cameras or confocal/non-confocal laser scanners. RNAs extracted from two different samples can also be simultaneously analysed with a microarray by labelling each with a different fluorescent label using either Cy-3 or Cy-5. At RIKILT-DLO we are currently testing whether the analysis of differential gene expression using DNA microarrays is an informative strategy for the safety evaluation of GM plants.

To address this question, cDNA libraries are constructed and enriched by a subtractive cloning approach for cDNAs preferentially expressed in either green or red tomato fruit. These subtracted cDNAs, as well as control cDNAs representing known tomato genes (*i.e.* sequence information derived from public sequence databases), have been spotted on arrays and are used for comparison of gene expression profiling of control and genetically modified tomatoes. In the near future we will construct more defined arrays containing genes of toxicologically relevant pathways, such as those involved in the synthesis of natural plant toxins (*e.g.* the glycoalkaloids á-tomatine and á-solanine).

mRNA fingerprinting

A very promising strategy to evaluate the equivalence of genetically modified crops is gene expression profiling. Conventional methods for the analysis of differential gene expression include Northern blotting, S1 nuclease protection, comparative expressed sequence tag (EST) sequencing, differential display and serial analysis of gene expression. More recently,

Kok *et al.* have designed a method of detecting altered gene expression by means of mRNA fingerprinting or reverse transcription polymerase chain reaction (RT-PCR). Based on the original concept of Liang and Pardee specific subsets of the mRNA population of a tomato plant were amplified. This enables the expression of different genes in the GM crop and genes in the parental line to be compared and any empirical information about significant alterations to be established.

The method is still under development and requires further validation, but has been shown to be reproducible and able to detect important differences in gene expression. Some of the most significant remaining problems are that the method largely depends on individual skills based on experience and is time-consuming, even in its most simplified form.

Proteome profiling

The introduction of Bt genes of bacterial origin may theoretically influence the nature of post-translation modification (*e.g.* glycosylation) in the transgenic tomato fruit. This is because as well as the impact of the protein moiety, for example, the type of glycosylation pattern may alter due to inactivation of glycosidases and glycosyltransferases in the host cell. In addition, glycosylation of the inserted Bt gene product in the plant might result in altered protein activity and stability.

The importance of structurally elucidating these oligosaccharides, especially the *N*-glycans, can be attributed to their widespread functions in the cell. These functions include correct folding, biological activity and stability of proteins as well as involvement in plant development. Moreover, *N*-glycans represent antigenic epitopes by themselves. Immunological as well as biochemical studies revealed the allergenic potency for the á1,3-fucosylation of the proximal *N*-acetylglucosamine residue. Additionally, also the β1,2-xylosylation of the β core mannose is suspected to be an allergenic epitope. Recently, there is increasing evidence that these *N*-linked oligosaccharides may have a substantial impact on the immunogenicity of glycoproteins, as *N*-glycans are suspected to play a certain role in the adverse food reactions of hypersensitized patients. At the Institute for Agrobiotechnology, Austria, this issue was tackled by developing a generic technology for isolating, purifying and characterizing *N*-linked oligosaccharides in plant tissues.

Our low-temperature acetone powder method appeared to be most suitable for the isolation of (glyco-)proteins from plant tissue (*e.g.* tomato fruit). The resulting (glyco-)protein fraction was proteolytically digested and corresponding glycopeptides purified by cation-exchange chromatography. The oligosaccharides (*N*-glycans including á1,3-fucosylated oligosaccharides) were released from the peptide backbones by *N*-glycosidase A (PNGase A) and fluorescently labelled with 2-aminopyridine.

Subsequently, pyridylaminated glycans were separated and analysed structurally by repeated two-dimensional high-performance liquid chromatography (2D HPLC) using a size-fractionation and a reverse-phase column in combination with an exoglycosidase treatment. MALDI-TOF (matrix-assisted laser desorption ionization time-of-light) mass spectrometry was applied for further confirmation of the observed structures. Sixteen different *N*-glycosidic structures could be detected in tomato fruit. The two most abundant glycans in all the tomato samples showed identical properties to those of the major *N*-linked oligosaccharides of horseradish peroxidase (MMXF: Man(3)Xyl-FucGlcNAc(2)) and pineapple stem bromelain (MOXF: Man(2)Xyl-FucGlcNAc(2)), respectively, and accounted for about 65-78 per cent of the total glycan content. Oligomannosidic glycans occurred in only small quantities (3-9 per cent). The majority of the *N*-glycans found was β 1,2-xylosylated and carried an á1,3-fucose residue linked to terminal N-acetylglucosamine.

A structural element was identified and shown to be an IgE-reactive determinant, which contributes to cross-reactivities among non-related glycoproteins, and is present in a wide variety of plants extracts. It is proposed that this structural element might also be of importance in some cases of food and pollen allergy.

The carbohydrate profiling of, for example, populations of green and red-ripe fruit of non-modified and modified Bt tomato plants (expressing the *cryIA5* gene) revealed no differences in the nature of *N*-glycans, nor were there significant differences in the relative amounts of glycan. Interestingly, the relative amounts found were quite similar across seasons and comparable to the commercial variety Notoro used as an extended control. As far as the *N*-linked oligosaccharides were involved it was not possible to detect significant alterations in the á1,3-fucose content of Bt tomatoes due to genetic modification.

Profiling of the metabolome

Among other factors, identity of a complex GM crop plant with its traditionally bred parent entails metabolism and the level of (un-)desirable substances contained therein. Therefore, we would like to build up an overall picture of components associated with different metabolic capabilities in the (non-)modified plant tissue. A multi-compositional analysis of the so-called metabolome may overcome the test limitations of, for example, single critical nutrient or key toxicant analyses and problems related to their limits of natural diversity. An attractive alternative is the utility of high-resolution proton (^{1}H) nuclear magnetic resonance (NMR) spectroscopy (^{1}H-NMR) in combination with separation techniques (liquid chromatography (LC)). Recently, we have shown that the approach of chemical fingerprinting using off-line LC-NMR can obtain information on possible changes in complex plant matrices due to environmental effects as long as GLP-like conditions for sample handling, data acquisition and automation of data handling are maintained. In principle, it will be possible to screen all the different low molecular weight components (MW <10 kDa) present in the metabolome of the plant tissue.

Toxicological profiling

The significance of detected compositional alterations may need to be further explored by *in vitro* toxicity assays in order to screen for potential adverse effects and for mechanistic studies. It may even be that the latter approach leads to a redesign, refinement or even replacement of (sub-)chronic rodent feeding trials. *In vitro* systems derived from organs and tissues from animals and humans, and various types of cultured recombinant cell lines, are successfully used for screening for the toxic potential of single compounds. However, their ability to screen for the potential toxicity of whole foods or extracts thereof has been insufficiently explored up till now, and has been examined by us within the framework of a tiered safety evaluation of GM plants.

The validity of using *in vitro* test assays for comparative testing of whole food products, food ingredients and extracts thereof depends on the sensitivity and selectivity of the toxic response induced upon exposure. However, since these models are not yet validated, they should be considered as early warning or alert-systems to identify potential changes in toxicity of the engineered product only.

genotoxicity endpoints

The mucosa of the gastrointestinal tract is clearly a primary, potential site of action for the effects of novel foods. Therefore effort has been concentrated on the development of a battery of intestinal cell systems (*e.g.* IEC-6, IEC-18, Caco-2 and INT407 cell lines) with different endpoints for initial cyto- and genotoxicity testing. The choice of a suitable method of sample preparation was considered of utmost importance in assaying vegetables.

Freeze-dried tomato fruit was extracted (10 per cent suspensions, w/v) in water, 0.9 per cent saline or chloroform:methanol (2:1, v/v) immediately before incorporation into the culture medium. It was encouraging that aqueous and chloroform/methanol extracts of, for example, red-ripe tomatoes up to a maximum of 10 per cent suspensions (w/v), had little or no general toxicity once the pH (> 6.0) and osmotic pressure had been taken into account. No toxicity to the intestinal cells Caco-2 and IEC-18 under short-term conditions was detected in red-ripe, non- and antisense RNA exogalactanase tomato using alpha-glutathione S-transferase (á-GST) lactate dehydroyrenase (LDH) leakage, reduction of 3-(4,5-dimethylthiazol-2-yl)-2,5 diphenyltetrazo-lium bromide (MTT conversion), neutral red (NR) uptake and total cellular protein as *in vitro* endpoints.

On the other hand, green tomato fruit displayed a pronounced toxic and cell-damaging effect. No statistical significant differences were found between transformants and isogenic control lines in terms of their activity in the assays. Treatment of tomato tissue with pepsin (ratio pepsin to tomato protein 1:250, w/w) did not result in an increase of or loss of inherent cytotoxicity. Supplementary investigations with naturally occurring plant components, for example, quercetin, β -carotene, á-lycopene, genistein showed that only steroid glycoalkaloids, such as á-tomatine, exhibited a rapid toxic effect, whereas á-tomatidine was ineffective. It was noted that the *in vitro* models did show general matrix-related toxic effects.

For example, á-tomatine diluted in a matrix of extracted red-ripe tomato tissue (final concentration of 5-10 per cent w/v) showed an increase of dose-response-dependent cytotoxicity in human Caco-2 cells (*e.g.* IC50 7.9 ± 2.1 mg/L versus IC50 33.9 ± 5.3 mg/L of á-tomatine only). Using the COMET assay, suspensions of freeze-dried tomato fruit up to a maximum of 10 per cent (w/v) did not exhibit DNA-damaging effects in a range of gastrointestinal cells derived from rats (IEC-6 and IEC-18) and humans (Caco-2, INT407). Genetic modification did not result in increased genotoxicity, but fruit extracts were able to suppress the DNA-damaging effects of the known genotoxins H_2 O_2 and N-methyl-N2 -nitro-N-nitrosoguanidine (MNNG).

Although these endpoints are important toxicological measures they do not indicate the molecular mechanisms of toxic damage. So far, pronounced adverse effects could only be observed with glycoalkaloids. The screening for subtle differences in toxic responses of other plant constituents appeared to be limited. However, therefore, the use of critical endpoints for such studies other than cytotoxicity should be clearly defined.

Assessing transcription factors

There is a demand for *in vitro* models that not only determine the (geno-)toxic potency of plant components or complex mixtures, but also provide information on the mechanism by which that compound exerts its toxic effect. As an *in vitro* profiling system the eukaryotic stress gene assay (*i.e.* CAT-Tox(L)) was assessed.

This utilizes human HepG2 cells stably transfected with chloramphenicol acetyltransferase (CAT) reporter constructs. The aim of the CAT-Tox(L) assay is to collect and quantify molecular responses to cellular stress and toxicity that have a transcriptional component in their regulation. The assay employs an ELISA method to measure the transcriptional activity of the promoter or response element constructs.

The advantage is that many of the stress responses occur before any measurable cytotoxicity, thus allowing monitoring of stress pathways at sublethal levels. Since 14 unique transcription responses can be measured simultaneously, the CAT-Tox(L) provides comprehensive cellular stress profiles.

The assay generates specific stress gene induction profiles for a variety of stressful and toxic plant components such as genistein, diadzein and sinigrin. Aqueous extracts of red-ripe, non-and antisense RNA exogalactanase tomato fruit dosed up to 0.05 g per assay did not cause any cytotoxicity to the HepG2 transfectants nor were there molecular responses to cellular stress or toxicity.

Extracts of non-modified and modified green tomato fruit showed cytotoxicity and induced the construct containing the xenobiotic response element (XRE). This gene-induction profile was not comparable to that of, for example, the reference molecules dioxin, 3-methylcholanthrene, benzo[a]pyrene or β -napthoflavone. These polycyclic aromatic hydrocarbons were characterized with a co-induction of cytochrome P450 1A1 and glutathione-*S*-transferase *Ya* subunit- CAT fusion constructs.

There was substantial evidence that the systemic fungicides fenarimol and bupirimate did not induce any of the genes in this assay. There was no promotor in the assay that could be linked to universal cellular stress. A-tomatine caused severe cytotoxicity at high concentrations and yet did not induce any of the genes (*i.e.* causing membrane damage). However, it was observed that á-tomatine diluted in a matrix of red-ripe tomato tissue (final extract concentration 5 per cent w/v) elicited an increased cytotoxicity concomitant with a dose-dependent response profile involving XHF (collagenase involved in inflammatory reactions sensitive to mitogens and cytokine IL-1).

So far, the approach of the CAT-Tox(L) model shows potential but there is a need for validation based on molecular mechanisms of plant constituents. Although the induction of these stress genes represents cellular perturbations, the mechanistic validations need to be extended before any conclusions can be drawn on the value of these findings. Certainly, the development of a database to explore relationships between a wide variety of compounds stressful to plants and meaningful stress gene profiles may serve as molecular fingerprints unique for a particular component.

Chemical fingerprinting

The principles for establishing a chemical fingerprint were directed towards the detection of alterations in compositions in ^{1}H-NMR spectra arriving from two populations, for example, non-modified counterpart(s) and an engineered one. In this way the unprocessed parts of the plant used for human consumption were fractionated and analysed, *i.e.* in tomato comparison on the fruit level. Genetically modified tomato varieties, such as the antisense RNA exogalactanase fruit, have been studied using this innovative technology.

The spread of measurements of the individual profiles was comparable for the GMO and its control(s). By subtraction of the ^{1}H-NMR spectra the differences of the means were calculated at 99 per cent confidence intervals between the GM crop average (n = 8/batch per line) and its isogenic control average (n = 8/batch per line). The difference of the means was defined here as all differences of the means of the amplitudes of individual NMR signals obtained from spectra of various populations.

The differences of the means were used to quantify how the transgenic crop plant differed from its parental line. The normalized 99 per cent confidence intervals of the normalized means of fractions A-E after eightfold independent replicates demonstrated that differences in amplitudes exceeding at least 20 per cent could normally be detected. This is an adequate sensitivity to allow statistical evaluation, and is in line with recommendations of the Nordic Council.

They proposed that if the average value of a parameter differs by more than 20 per cent an explanation should be sought. With the purpose of establishing those differences that may arise from metabolic effects linked to genetic modification the default statistical analysis used a mixed effects randomized block model. Random block effects such as variations in the stage of cultivation, location, logistics or climate had a significant impact on the overall chemical fingerprints of cultivars.

For example, chemical fingerprints varied much more between field sites and seasons than between replicates in one plot. A tomato background mean (*i.e.* crop mean value) was established by constructing an extended range of non-modified tomato references for comparison with natural variability, and to assess the impact of external factors (*i.e.* variations in season, climate, etc.). The crop mean value was used to quantify how the differences of the means between the transgenic crop plant and its control differed from the natural diversity of commercial counterparts.

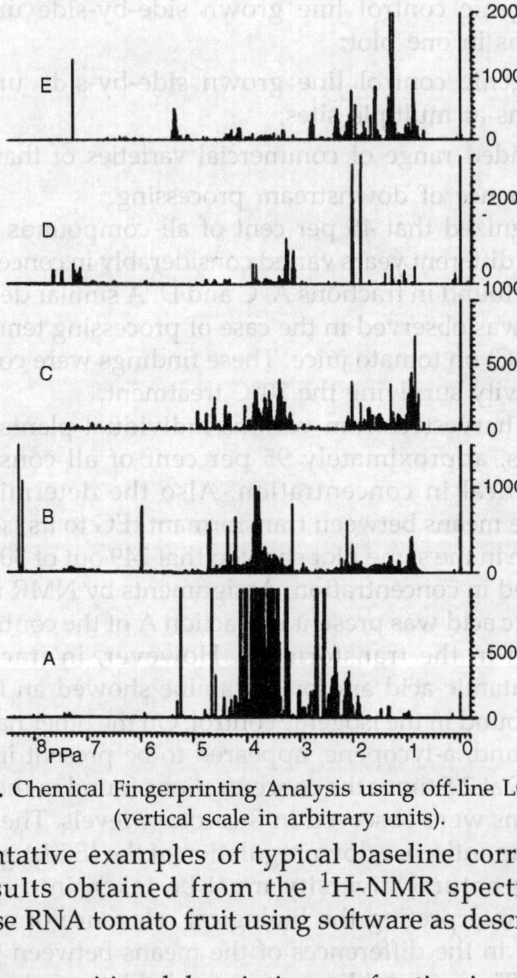

Fig. Chemical Fingerprinting Analysis using off-line LC-NMR (vertical scale in arbitrary units).

Representative examples of typical baseline correction and peak selection results obtained from the ^{1}H-NMR spectra of antisense-exogalactanase RNA tomato fruit using software as described in Lommen *et al.*.

A global compositional description per fraction is as follows: fraction A, polar compounds such as fruit acids, glucose and other monomeric and oligomeric sugars, amino acids and TCA-cycle substrates; fraction B, large variety of aromatic, aliphatic and sugar-like compounds; fraction C, primarily non-tomatine glycoalkaloids and aromatic compounds; fraction D, tomatines and aromatic/indole-like compounds and fraction E, primarily lipids and carotenoids. Typically the detection limits in the various fractions are estimated in the range of mg/kg wet weight tomato tissue.

To differentiate between compositional changes either due to genetic modification, genetic variability or environmental variations the chemical fingerprints were studied in a hierarchical approach. Hence, the fingerprints of the GMO are compared to those of:

- The isogenic control line grown side-by-side under identical conditions in one plot;
- The isogenic control line grown side-by-side under identical conditions at multiple sites;
- An extended range of commercial varieties of that crop;
- The influence of downstream processing.

It was recognized that 45 per cent of all compounds of the control line harvested in different years varied considerably in concentration. Most differences were found in fractions A, C and D. A similar degree of impact on composition was observed in the case of processing temperatures (*e.g.* 100°C versus 70°C) on tomato juice. These findings were compatible with the protease activity surviving the 70°C treatment.

But in fruit harvested from various individual plants, divided into two populations, approximately 95 per cent of all constituents were practically identical in concentration. Also the determination of the differences of the means between transformant tEG to its isogenic control bred side-by-side in the same plot showed that 249 out of 3000 amplitudes (8 per cent) varied in concentration. Assignments by NMR indicated that, for example, citric acid was present in fraction A of the control at 1.4 times the levels found in the transformant. However, in fraction A of the transformant glutamic acid and/or glutamine showed an increase of 1.3 times the levels found in the isogenic control. On the other hand, one single aromatic compound, á-lycopene, appeared to be present in fraction E of transformant tEG at 2.5 times the concentrations found in the parental line.

No alterations were observed in ß-carotene levels. These effects may be the result of the antisense downregulation of the *tEG1A* gene in tomato. In our view the natural diversity must be taken into consideration, however, when interpreting the biological relevance of any significant difference found in the differences of the means between the GMO and its parental line. The parental line was extended by entering non-isogenic commercial references (*i.e.* crop mean value). Although limited in the range of lines, the comparison of the crop mean value to the differences of the means showed that the contribution and magnitude of differences (*e.g.* citric acid, glutamic acid/glutamine, á-lycopene) diluted out. It was recognized that in the agricultural practice the conditions of culturing, processing and storage of appropriate references will be mostly unknown and they may have a non-isogenic genotype.

Notwithstanding these complicating random block effects we recommend that, for example, the chemical fingerprint should not be limited to the transgene and its isogenic parent bred under identical circumstances only, but should also include a sampling through multiple sites and a comparison with an extended range of controls of that crop. Otherwise it would be a too limited approach towards the establishment of the claim of substantial equivalence and it would neglect the ranges of natural variations in that crop. In future, if many more lines have to be compared with each other this approach may benefit from multivariate statistical methods.

Gene expression profiling

A second potential application of DNA microarray technology in safety evaluations concerns screening for biological effects of (fractions of) plant compounds that have been found by other profiling techniques to be present or modified in the genetically modified crop plant when compared to the parental line and other references. At RIKILT-DLO we are testing the feasibility of this strategy by analysing defined food compounds for possible effects on human intestinal gene expression.

To this end, human intestinal cell lines are exposed for different time periods to various amounts of crop plant components and the mRNA isolated. The labelled mRNA is hybridized to DNA microarrays containing unknown intestine-specific genes, as well as already identified control genes known to be intestine specific or to be functional indicators (for toxicological processes, apoptosis, etc.). As a source for unknown, intestine-specific genes we use cDNA libraries from human intestine biopsies and human intestinal cell lines which have been enriched by a subtractive cloning approach.

Having established the usefulness of this approach for the identification of biological functions of defined plant compounds, these cDNA microarrays will be exploited to screen, for example, GM crop plants for possible effects on human health. The concept of substantial equivalence is broadly accepted as a basis for the risk assessment of GM crops and derived novel foods. Data requirements for establishing substantial equivalence should be based on identifying unintended metabolic perturbations in the engineered crops that have been caused by genetic modification. A generic approach cannot be envisaged, however. In all cases the collation of data should be directed by the type of genetic modification and related consequences. It is recommended that unintended effects should be identified on a case-by-case basis, since such effects are dependent on the genotypic and phenotypic parameters of the new gene products involved. The application of innovations in molecular genetics research will help to define the conditions under which the new food products can be marketed. The present approach has generated a sound scientific basis for the evaluation of the potential of unintended effects in GM crops and derived food products. The results may assist regulators and legislators to test and, if necessary, to improve the applicable regulations. Moreover, the technologies described and the results obtained may serve as a general framework for the risk assessment of GM crop plants, and may contribute to a better understanding by the public of recombinant DNA techniques in plant breeding and their implications.

It is concluded that a parallel approach focusing on a comparative analysis using informative profiles of molecules at various integration levels (*e.g.* mRNA, protein, metabolites) including a toxicological profiling of relevant plant matrices, offers good prospects for the identification of hazards related to unintended effects. The choice of comparators and of external parameters to support a claim that a GM crop plant presents no unintended effects compared with its traditional counterpart should be based on sound scientific judgement. It is recommended that data are generated on the unprocessed part of the plant, such as the tomato fruit or potato tuber, used for human and/or animal consumption. Furthermore, data banks should be set up containing information on natural variations of essential plant or other product constituents (*i.e.* crop mean values).

The chemical fingerprinting technique, using a combination of off-line liquid chromatography and proton-NMR imaging, appears to be a powerful screening method for the detection of secondary effects in the metabolome that may be applied on a routine basis. The carbohydrate profiling technique is useful for the detection of unexpected changes at *N*-glycan levels; in particular it is applicable to the detection of post-translational modifications in newly expressed proteins as well as in the whole GM crop plant. DNA microarray technology is rapidly evolving and much effort is being put into improving spot density, reduced production time, and increased reproducibility and sensitivity.

Improving the latter is critical because it makes it possible to use smaller amounts of starting material. It will depend on technical aspects such as the quality of the scanners and spotting machines, the development of fluorescent dyes with improved characteristics (*e.g.* narrow excitation and emission peaks; high level of photon emission; resistance to photo-bleaching), supports with reduced background and more target sequence binding capacity. DNA microarrays generate a huge amount of complex hybridization data and major challenges are to develop more advanced computer software to help to find statistically significant correlations within, and between different experiments, and to link the data to sequence information and submitted expression profiles available in public databases. Application of these technologies in combination with other measurements of, for instance, the performance and quality may limit or even replace animal feeding studies aimed at the detection of unintended effects.

Together, these techniques, known as functional genomics, will enable us to answer questions about what happens to expression and protein composition when a plant cell changes metabolic state, for example, due to genetic modification or environmental factors.

It is foreseen that this holistic view on the cellular machinery will be achieved soon, as the coincidence of genome sequencing with improvements in the analysis of expressed proteins is of growing importance and continues to be successful. Eventually, microarrays for genomic studies will probably also be used to evaluate the numerous constituents of GM crop plants. Some constituents might not cause obvious changes in cellular behaviour or morphologic appearance but could cause subtle metabolic alterations that would show up when the mRNA content was interrogated by an array.

It is likely that microarrays might be used to evaluate the life cycle of a plant much more precisely and to understand the complex metabolic control systems of plants.

The ultimate test for GM crop plants will be acceptance by the public in large. It is therefore extremely important that industry, the scientific community and the regulatory authorities are able to respond to queries from the public by providing transparent information on the criteria for safety assessment. Improved safety assessment at the molecular level should refine and complement strategies by anticipating the potential risks and sources of unintended effects in GM plants.

Virus Resistant Crops

Some viruses infect people or animals and other viruses infect plants. Plant viruses reduce the productivity of annual crops and can kill fruit trees. Some plant viruses are spread by insects. Plants can be protected from those viruses by using insecticides or other pest management methods. There is essentially nothing else that a farmer can do to protect his crop from virus damage, except to grow a different crop. But genetic engineering a plant to protect it from a particular kind of virus is quite easy.

A gene from the virus which encodes a protein in the virus' outer coat is copied into the plant's DNA. The plant then makes the coat protein, which is harmless, but which stimulates the plant's natural defences. Virus resistance traits have been introduced into many crops, including squashes, tomatoes, potatoes, tobacco and, perhaps most dramatically papaya. Recently, cultivated Hawaiian papayas were hit by a devastating virus which essentially eradicated the commercial variety. Only the virus resistant transgenic papayas survived. If you like papaya, you can only buy the transgenic variety. Nobody can grow any other kind.

The Potato Famine

In 1840s Ireland, the potato crop was devastated by a late blight fungus (Phytophthora infestans) and Irish people starved en masse. That fungus could reappear at any time in any place and wipe out a potato crop. Some varieties of potato have previously had some resistance to late blight fungus, but now a fungal strain has appeared in Russia that destroys those previously resistant varieties.

This year (2001) a similar fungus appeared in potato fields in Prince Edward Island and 630 million pounds of potatoes, the island's principal crop, had to be destroyed. But very recently, scientists were able to transfer a gene from alfalfa to a potato plant and the resulting potato plant is able to resist the fungus and thrive. Potatoes also rot. A principal cause of potato rot is the bacterium Erwina carotovora, which has been called the flesh eating bacteria of the plant kingdom. Now a gene that confers resistance to E. carotovora has been coupled to a control gene that turns on when a plant has been wounded, and this construct has been transferred to experimental potatoes. As the researchers hoped, the modified potatoes, when punctured by a toothpick and exposed to E. carotovora, had almost twenty times less rot than unmodified potatoes.

Sentinel Crops

A recent innovation is a plant intended not for food but for quality control. It contains a gene derived from a luminescent jellyfish, but in all other ways it is identical to the food crop it is planted alongside. When these sentinel plants experience a lack of water, they literally glow in the dark. The farmer then knows that his crop must be watered or whether irrigation can be postponed. In the western U.S. water is scarce. Agriculture is the biggest user of water. Wasting water is intolerable. For example, so much water is taken from the Colorado River for irrigation that the river flows into Mexico a mere trickle, and it never gets to the sea at all. So this is yet another way that transgenic crops can benefit the environment.

Building with Silk

Silk is composed of two proteins, fibroin and sericin. The gene for fibroin has been transferred from a silkworm to a goat, and is expressed as a component of its milk. Soon we may also expect sericin to be transferred. It still remains to be seen whether technology can be developed to spin these proteins into a fibre. That has already been accomplished with the kind of silk spiders use to make their webs. Genes for the two spider silk proteins were transferred to cells cultured from cow udders. Those cells then made the proteins. Happily, the spider silk can be spun by forcing a solution of its two protein components through a tiny nozzle. The proteins self-assemble into spider silk strands. The same genes have since been transferred to live goats and when those goats are old enough to produce milk, it should be possible to make large quantities of spider silk very cheaply. Silk is an extraordinarily strong material, stronger than steel. In the future we may be getting our strongest building material from a farm instead of from a mine.

Safer Meat

Escherichia coli are friendly bacteria that live in our intestines and contribute to our health. But there is one strain of E. Coli (designated as O157:H7) that can make us sick, even kill us. We can get it from inadequately cooked meat. The E. coli infected meat comes from cattle with the virulent E. coli strain in their intestines. A cow's digestive system is adapted to digesting hay and grasses. The food first goes into a pre-stomach called a rumen. That's why cows are called ruminants. In the rumen, microorganisms turn the indigestible cellulose into nutrients the cow can assimilate. The food is then passed to a true stomach, and finally gets to the cow's intestines, where E. coli can live.

Therefore to guarantee against the virulent strain thriving in the cow's intestine, one needs to get some sort of prophylactic agent into the intestine. Antibiotics won't do. They would kill the cow's normal intestinal bacteria, and besides, it isn't a good idea to overuse antibiotics. There are antibodies specific to the virulent strain of E. coli, but they would be destroyed by passage through the cow's stomach before reaching its intestines. Genetic engineers are working on a neat solution.

They are developing a transgenic animal feed which resists complete digestion in the stomach and delivers the antibody, specific to the virulent E. coli strain, into the intestines. There are over 60,000 cases of E. coli illness in the United States each year. There would be many more except for an extensive programme of meat inspection. Even this understates the problem, because meat, if contaminated, has to be destroyed.

E. coli infections were not such a serious problem when cattle were raised exclusively on grass and hay. On that diet, there isn't much digestion going on in the cow's intestine and the E. coli populations are comparatively low. But today's cattle spend the last weeks of their lives in feed lots, being fattened up on grain, which is digested in their intestines, leading to much higher populations of E. coli. So another way to solve the E. coli problem would be to raise leaner cattle and skip the feed lots.

Reduced Need for Fertilizers

One of the ways that farmers get better yields is by providing their plants with sources of organically bound nitrogen and phosphorus. These can be provided either by applying chemical nitrates or phosphates, or by using manures or decaying vegetation as sources of the same nutrients. Nitrogen is the largest constituent of the atmosphere, about 80 per cent. It may seem paradoxical that unfertilized plants could suffer from a nitrogen deficiency while immersed in a sea of nitrogen gas, but it is just not available in the form they need. Of living things, only certain bacteria (and human chemists) have evolved the means to convert nitrogen from the atmosphere to a form useful to plants. But some plants, primarily legumes (peas and beans), have a symbiosis with these nitrogen fixing bacteria. The plants provide nodules on their roots that protect the nitrogen fixing bacteria, which then enrich the soil around those roots.

Not only does this permit the legumes to grow luxuriantly without nitrate fertilization, but it makes the soil fertile for other plants growing in the same soil later. The technique of crop rotation is one of the oldest techniques of agriculture. Scientists hope to be able to transfer the genes which direct the formation of the nodules to other crops. If this is successful, the need for fertilizers would be dramatically reduced. Unlike nitrogen, phosphorus is not a constituent of the atmosphere. There is no short-term likelihood that scientists will find a genetic engineering way to replace fertilizers that provide phosphates.

The best hope for phosphate replacement would be to breed or engineer plants that make more efficient use of the phosphate available to them. If it proves impossible to engineer plants for nitrogen fixation, there are still options which can let them use fertilizers more efficiently. An enzyme called glucine dehydrogenase is involved in utilization of fertilizers. The gene for glucine dehydrogenase is present in most crops, but it is expressed at low levels, because the control genes turn it off more than on. A genetic transformation of wheat which promoted increased synthesis of glucine dehydrogenase was 29 per cent more effective in utilizing the same amount of fertilizer as the unmodified variety. The increased efficiency can either be used to grow more crop on the same land, or to cut down on the need for fertilizer to grow the same amount of crop.

Crossing over in Plants

Plants and animals have two of each gene, on separate but similar chromosomes, analogous to having two copies of the document. The two copies are usually slightly different. By analogy, if the document were The Lord's Prayer, one copy might read The Lord is my shepherd! I shall not want. The other copy might read God is my guide! He provides for me. In sexual reproduction, the two documents are sliced apart and put back together, so that they might read The Lord is my guide! He provides for me. and God is my shepherd! I shall not want. This process is called crossing over. This happens in the sex cells of both parents and then the offspring inherit one "document" from each parent. Crossing over is a major source of genetic variability. Sometimes, however, it produces the same kind of nonsense as in the example "don't be happy! drink the water!".

Transposons

Genes are arranged on the chromosome in order but many genes make a habit of jumping to another part of the chromosome. They are routinely snipped out of the DNA and put back in a different place. These jumping genes are called transposons. By being moved to a different part of the DNA, the functional expression of a transposon changes. A familiar example of a transposon is a gene that determines the colour of the kernels of Indian corn.

Viruses

Many viruses insert their own DNA into the host's DNA document, where it gets copied and becomes part of the genome. More than half of the DNA in the human genome originated as invading viruses. These viruses can carry useful genes from other species into the infected animal or plant. Many human genes originated in other species by this mechanism.

Mutations

With billions of letters in a DNA document, there are always a few copying errors, called mutations. A mutated gene has a changed meaning. It may specify a slightly different protein, or it may switch another gene on or off at a slightly different time. These changes are usually harmful, but tolerable, especially when we have two copies of a gene and one is unchanged. Occasionally a mutation is either deadly or beneficial. According to the theory of evolution, all of our genes originated, in the distant past, as mutations of other genes, a completely random process. The critics of genetic engineering mean to leave you with the impression that nature has evolved a finely tuned but fragile system of inheritance but that genetic engineers have no good idea what they are doing. But actually, the genetic engineers make only small well controlled modifications to a genome, whereas nature often makes large and random changes. The likelihood of unexpected effects in transgenic technology is small. Unexpected changes in conventional breeding are virtually certain.

Glyphosate

We earlier mentioned glyphosate, an environmentally benign herbicide. It is biodegradable, lasting approximately two days after use, and its molecules bind to soil so it does not wash into streams or enter groundwater. Even if some madman were to dump it into a waterway, it is 230 times less toxic than the herbicide it replaced. It has been found safe and approved for use in almost every country, including all the countries of the European Union. Yet the anti-transgenic crowd seeks to ban it in Europe using the pretext of uniformizing regulations from one European country to another. What reason is given for banning the safest of all herbicides? Two reasons. It is supposed to be 'implicated' in non-Hodgkins lymphoma, and it is alleged to harm beneficial insects. Examining the evidence for these two claims reveals how some opponents of GMOs will use anything, no matter how shaky, to achieve their purposes. Non-Hodgkins lymphoma patients were asked to recall what pesticides they had been exposed to in the past several years. A statistically insignificant number of them (four) mentioned glyphosate, not surprising since it is so widely used. Even though the investigators considered this association meaningless, it was enough for Greenpeace to demand a ban. The claim about harming insects is even weaker. Several species were exposed to glyphosate and the control group were not exposed. There was similar mortality in both groups.

But there was mortality. This is enough for Greenpeace to claim that glyphosate kills beneficial insects. Greenpeace ignores a follow-up study by the International Organization for Biological Control, which concluded that glyphosate was exceptionally safe. One might have expected the IOBC to be hoping that it would find something wrong with glyphosate, since its charter is to popularize biological controls. But IOBC's scientists are ethical. The real reason for banning glyphosate is that it is made by Monsanto and used as the herbicide for the herbicide-resistant transgenic soybeans.

Labelling Transgenic Food

Once the opponents instigated a doubt about the safety of genetic engineered foods, and remember that there has not, in ten years, been even the slightest evidence of that, the next step was to agitate for labelling. Why, after all, shouldn't consumers have a choice? I agree people who want to avoid transgenic food should be able to make that choice. But labelling can take two forms. One could label food which is not transgenic, or one could label food which is transgenic. I can't imagine anyone objecting to labelling foods which are transgenic-free. But the anti-transgenic demand is adamant for the other choice. We can say that the difference is between those who would label the non-transgenic food with a smiley face and those who would label transgenic food with a skull and crossbones. Labelling sounds so reasonable. How could anyone oppose it if they didn't have something to hide? But labelling has a down side. It costs.

The cost is not in the ink to print the label. It is in keeping the products separate. We are talking now mostly about soybeans and corn. These grains are harvested by the hundreds of tons, shipped in railroad cars, stored in grain silos, sold in futures contracts. Approximately 70 per cent of processed foods in the US have some ingredient derived from biotechnology.

This includes most milk and cheese, sugar, soy products, corn and corn sweeteners, vegetable oils, etc. 63 per cent of US grown soybeans are now transgenic. Neither buyers nor sellers distinguish between transgenic and conventional soybeans. Your tofu is a mixture of both kinds. (In my local supermarket, there is only one brand of tofu available and the manufacturer has recently decided to use only organic soybeans, which are specifically labelled non-GMO. At the same time, the package size changed from sixteen to fifteen ounces and the price was raised by $.25.) To keep the two kinds separate, we would need, at a minimum, to have two separate distribution channels, two storage systems, two futures markets. In my opinion, the pressure for skull and crossbones labelling is really a pressure to increase the cost of the genetic engineered food. So far, the transgenic food has had only producer advantages — it is cheaper to produce. Take away its cost advantage and it is no better than the conventional foods.

Never mind that corn with the Bt trait, one third of the US crop, is grown without pesticides. Never mind that the transgenic soybeans prevent soil loss and global warming. The Economist magazine estimated that segregation of grains would add a 25 per cent premium to the price of some processed foods, like packaged cereals. This is probably an over-estimate, but we should remember that these costs will fall inordinately on people with low income. But if this were the only problem with labelling, the food industry giants would quickly adopt grain segregation and labelling. They are being forced into grain segregation anyway by the labelling regulations of nations that import US grain.

The industry is really concerned that the labels will make it easier for groups like Greenpeace to boycott their products. Stop the scare campaigns and the resistance to labelling will disappear overnight. If you doubt that demands for skull and crossbones labelling can be used to purposely create a disadvantage, imagine what your reaction would be if someone were to propose that products assembled by hispanic workers must be so labelled. The public demand for labelling would be justified by a "right to information", but just below the radar screen you would not be surprised to hear that hispanic workers might be illiterate, or might be illegal immigrants, or might be drug users. But you would recognize that the demand for labelling was meant to disadvantage hispanics. We have established a precedent in this country about labelling. The government does not mandate labelling without a very good reason. The main exception is when safety (skull and crossbones label) is involved. Once the government decides that, say, a certain chemical pesticide is safe, nobody can require the food grown with that pesticide to be labelled. When a segment of the consuming public wants a label (smiley face) about a trait that it cares about, the market provides such a label and it is reflected in the cost of the product.

For example, there are people who prefer to eat food grown with no chemical pesticides. They buy "organic" food. Everyone understands that the label "organic" means that the food was grown without pesticides. Everyone also knows that they have to pay more for the food labelled organic. Similarly, orthodox Jews have certain religious rules about what they eat. These include a requirement that cows and chickens must be slaughtered in a particular way. This is a requirement for the Kosher label. Jews do not expect all meat to have a label detailing how the animal was slaughtered. Even in Israel, the people who want the label pay for the privilege, and kosher meat is often quite a bit more expensive that ordinary meat.

Environmental Scare Stories

We mentioned earlier a concern that some genetic engineered crops could cross breed with wild relatives. To the pseudo-environmental groups (those who like to call themselves green even though much of what they advocate would harm the environment), this absolutely rules out any permissible use. Their argument goes as follows: "We don't have any idea what plants could cross breed with what other plants, and we don't have any idea what effect the unusual genes would have in the wild plants. Therefore we should take no chances." In fact, hybridization between different species is extremely uncommon and it has never been observed between distantly related species. A tomato won't cross with a potato, even though they are both members of the nightshade family.

Yet the GE opponents would have us believe that just about any plant can hybridize with any other. And they have quotes from PhDs to prove it. Actually there is exactly one controlled experiment showing a hybridization between a genetic engineered crop and a wild species. The crop was a transgenic canola (oilseed), with that same herbicide tolerance gene we have encountered before. Here is what PhD Jeremy Bartlett, of the John Innes Plant Research Centre of Norfolk, England, wrote to the Manchester Guardian. He said we don't know what plants will hybridize with what other plants. He said that there is a documented example of transgenic canola hybridizing with a wild mustard. He went on in the same letter to talk about the possibility of gene transfer to soil organisms. Dr. Bartlett is a PhD in biology. He surely knows that there is little basis for speculating that plants will pass genes on to soil organisms. But let's give him the benefit of the doubt where a speculation is concerned. Still he must have known that the documented case he was referring to was observed at a test plot of the John Innes Research Centre, his own institution.

The transgenic canola was planted in the centre of the plot and various other species were planted at various distances to measure the rates of hybridization. There was only one case observed. The canola hybridized with the wild mustard. Jeremy Bartlett means for you to think that if an oilseed can hybridize with a wild mustard, then anything can hybridize with anything. Fortunately, the John Innes Centre puts its research reports on the world wide web, so I read the actual report. Guess what? Canola is a hybrid itself, a cross between Brassica rapa and Brassica napus, two closely related wild plants in the same genus. Brassica rapa is wild mustard! So we don't have evidence for a transgenic crop hybridizing with just about anything else imaginable, beyond the capability of man to anticipate. We have evidence of a plant crossbreeding with its closest relative, placed in the test plot because it was so likely to crossbreed, and it was the only case observed.

When the biotechnology opponents write or talk about this, they always say wild mustard and canola, never Brassica rapa and napus. Dr. Bartlett means to play with our minds, to mislead us. Hybridization with wild plants is a concern, but it is a managable concern.

5

The Alternative Agriculture

The Neglegence

This chapter makes the claim that, despite the emergence of alternatives, the trend towards genetic uniformity is likely to be perpetuated owing to the emphasis now placed upon new biotechniques. Biotechniques provide more powerful tools to achieve genetically uniform ideal plants, but they could also be deployed as part of strategies aimed at promoting deployment of *in situ* diversity. The argument presented here is that biotechniques are closing the door which alternative movements managed to prise open, if only briefly, in the 1980s. This could be made in various ways, but here, the argument is developed through analysis of the intellectual property rights (IPR) issue, since this has radically affected the issue of genetic resource control in recent years.

Support for biotechniques takes place against the backdrop of a global economy in which concern to maintain 'technological leadership' and 'competitiveness' has prevented serious consideration of important questions concerning technological options, and the need to consider the matching of ends with technological means. The more deep-seated this concern, the more likely it becomes that institutional change will underpin it, further legitimising the same concern. This is the domain of Arthur's self-reinforcing mechanisms. For this reason I suggest that we are becoming increasingly locked in to genetic uniformity only moments after it seemed that the door was possibly opening for (again, possibly) genetically more diverse alternatives.

Genetic resource control

When southern corn leaf blight hit the US maize crop in 1970, the solution was found not through re-introducing genetic diversity, but in genes from Mayorbala maize found in a West African field. West African germplasm was used at virtually no cost to save US farmers from a disease that had cost them $1 billion in 1970. That US farmers should have been assisted so freely owed itself to the system of free exchange of germplasm operating at the time, based on the principle that biodiversity was 'the common heritage' of humankind. The UN's Food and Agriculture Organisation (FAO) had discussed genetic resources at its founding conference in Quebec in 1946, and with the International Biological Programme (IBP) hosted a conference on plant genetics in 1961. Little was being done to address the issue of genetic erosion. Most collecting of germplasm was done by academics in universities, and virtually all was done at a sub-national level, an exception being that conducted on maize. The issue of crop germplasm conservation was highlighted by the work of Erna Bennett, who in 1967 organised the second IBP/FAO Conference on genetic resources and subsequently set up the Crop Ecology Unit at the FAO in Rome.

Only a year after southern leaf blight destroyed much of the US crop, a cold winter followed by a dry spring drastically affected the wheat crop of the Soviet Union. Much of the land was planted to the Besostaja variety which was neither cold tolerant nor drought resistant. Wheat prices shot up, and when famine hit the Sahel region in Africa, Malthusianism became fashionable once more. Against this backdrop, Sir Otto Frankel, Bennett's co-editor for a path-breaking book on the subject, persuaded the 1972 Stockholm Conference on the Human Environment to adopt a resolution calling for concerted action on genetic resources. The need to conserve genetic resources was at last being taken seriously as the advancing BCM mode hastened their disappearance.

Hambridge and Bressman recognised long ago the tensions between free exchange in germplasm, and unfree exchange of the product derived from it: From its rivals a nation may get the wheat germplasm that enables it to supply its own needs or overwhelm those rivals in international trade ... Will nations have the wisdom to deal with this situation, or will it lead to more bitter rivalries and more deadly conflicts, as the beneficent science of chemistry has enormously increased the deadliness of war? In his use of modern science, man has proved again and again that he is a bright child playing with fire.

The Wardian case, invented in 1829, facilitated an exodus of germplasm to one of a growing number of botanical gardens. The 'botanical chess game' has played an important role in shaping the international division of labour, a fact recognised by Marx: You perhaps believe, gentlemen, that the production of coffee and sugar is the natural destiny of the West Indies. Two centuries ago, nature, which does not trouble herself about commerce, had planted neither sugarcane nor coffee trees there.

Metropolitan powers appreciated that control over commodity trade depended on restrictions on the movement of germplasm. History is therefore replete with examples of the heroic efforts of plant explorers in overcoming embargoes on the movement of seeds, the breaking of which was in many cases punishable by death.

Only in the post-colonial world did 'free exchange' reach its truly international apogee. Some interpret this as allowing gene-poor northern countries in the developed world to maintain access to germplasm residing in the gene-rich, financially poor, countries of the developing south. Yet if, as Galeano writes, the international division of labour was organised 'not by the Holy Ghost but by men', so it was with the system of germplasm exchange. As such, it could be changed by men.

The Seed Industry

Unless they were crossed inbreds, farmers could save seed for planting the following year without appreciable yield loss. The first claims for plant patents were made in 1885. In 1922, lawyers met in London to discuss patent protection for plant varieties but no action followed. It was the nursery industry which was primarily responsible for the passage of the Plant Patent Act of 1930 in the United States. For nurserymen, the obstacle to proprietary ownership of varieties lay in competition from other nurserymen, not farmers, since trees could easily be propagated in competing nurseries.

The Plant Patent Act of 1930 made it possible for asexually reproduced plants to be patented, with the exceptions of potatoes and Jerusalem artichokes. The rhetoric used in support of the act, that breeding had made such significant advances over the past decades, was actually completely irrelevant as far as asexually reproduced plants were concerned. Most of the work done by nurserymen lay in multiplying varieties that had been discovered by chance, and that were the product of insect or wind pollination, raising issues as to whether they should have been eligible for patents.

Fowler concludes that: 'The PPA did not recognise the individual inventor or the creative act as much as it recognised and rewarded the system that produced the new variety, whether by luck or by designs'. In France, ever since the turn of the century, rose breeders had been seeking the same recognition as inventors of machines. Early attempts were rejected by lawyers on grounds that even full disclosure would not make it possible for breeders to reproduce a variety. By 1928, however, there existed in the Ministry of Agriculture *de facto* protection of breeders' rights through an 'identity and purity service'.

The Italian High Court declared plant varieties patentable in 1948, but confusion led to calls for a special plant patent law. By 1957, with the view that plants and animals should not be patented in the ascendancy, the International Association of Plant Breeders for the Protection of Plant Varieties accepted an invitation to host a conference in Paris on plant breeders' rights (PBR), leading to the establishment of the Union for the

Protection of New Varieties of Plants (UPOV) in 1961, whose International Convention was revised in 1972, 1978 and 1991.

For the most part, PBR legislation has been true to the UPOV Convention, requiring that plants pass the DUS test. The 1960s and 1970s saw several countries either joining UPOV or implementing a system of PBR of their own. Key to the passage of these acts was the belief that sexually reproduced plants could breed true, which, for sexually reproducing plants, is only the case for pure-line varieties (those having undergone four to nine generations of selfing). Also critical was the definition (thought by many to be impossible) of the term 'variety'.

The UPOV resorted to 'nothing other than a description of the steps of the method of breeding', or more accurately, pure-line (Mendelian) breeding. Hence, the extension of IPR to plants through PBR was an institutional innovation shaped by, and made possible by, changes in breeding techniques and technology respectively. However, these were not institutional changes waiting to be implemented as soon as these techniques emerged. Just as there were technical options open to breeders, so the institutional changes made represented a choice from myriad possibilities.

PBR facilitated an increasingly international outlook on the part of the seed industry. Modern varieties were spreading across the globe, and notwithstanding some efforts to improve disease resistance in new varieties, new seeds made increased use of other inputs more likely. In the 1980s, policy-related or structural adjustment lending undertaken by the World Bank advised privatisation of input supply industries, and an expanded role for private sector seed research and distribution, especially in the development of hybrids. The emphasis began to shift, as it had done in the developed world, away from the public system in favour of reduced public sector involvement.

As late as the 1960s, there were few multinational companies in the seed industry. A wave of acquisitions occurred in the 1970s as seed companies were bought up by transnational corporations, mainly food trading and petrochemical companies. Food traders, seeking to open up new export markets in the era of US 'food power', sought to extend their activities upstream.

The development of high-input seeds by international agricultural research centres had also led (agro-) chemical companies to seek new markets in the developing world, so these companies sought to market seeds through the same channels. With PBR legislation in place in many developed countries, seeds were no longer a weak link in the input supply industry. UPOV, by creating a degree of harmonisation in PBR legislation, fostered the emergence of a global seed industry, whilst the horizontal integration of agricultural input supply has deepened the inter-relatedness of inputs over time.

The Challenge to the BCM Paradigm

The BCM mode has come under, and continues to operate in the face of, considerable pressures for change. These are due to its:

- Ecological impact;
- Impact on food quality and health (in farming, and in consumption);
- Impact on rural communities;
- Being supported by state policies, and high levels of farm support, which effectively exacerbate the problems mentioned above.

It is beyond the scope of this book to address each of these in detail. Suffice to say that, in the words of Almas and Nygard, the BCM mode has produced 'some of its own executioners'.

Aims to reform agricultural and farm support policies seek, increasingly, to re-direct support towards environmentally sound practices. These are generally believed to imply reduced use of agrochemicals, and also seeds. To the extent that it continues to be allowed, seed-saving becomes more economically attractive in times of low output prices. In the spirit of challenges to the BCM mode, it is to the issue of seeds and germplasm that we now turn.

Sowing the Seeds of Discontent

As the Rockefeller and Ford Foundations expanded their efforts in international agricultural research, they began to seek public funds to support their work. Following a meeting in Belaggio in April 1969 organised by the Rockefeller Foundation, 15 governments attended the first meeting of the Consultative Group for International Agricultural Research (CGIAR). The World Bank would provide a secretariat and administer finances. International agricultural research, the year after Norman Borlaug received a Nobel Prize for his work on dwarf wheats in Mexico, had come of age.

In 1972, the CGIAR's Technical Assistance Committee (TAC) convened a meeting in Beltsville, USA, to formulate an international strategy for genetic resources conservation. After much debate, the International Board for Plant Genetic Resources (IBPGR) was set up as part of the CGIAR network. Under the IBPGR's system, the majority of the world's genebank accessions, mostly from developing countries given to believe that genetic resources were the common heritage of humankind, were stored in genebanks in developed nations increasingly predisposed to the notion of IPR over germplasm.

In the late 1970s, developing countries were increasingly concerned by IBPGR statistics showing that of more than 1.9 million samples stored, 55 per cent were in developed countries and another 14 per cent were held in the Northern (donor) dominated CGIAR system. Collections were clearly biased towards crops of interest to the developed countries, the top cereal crops representing more than 75 per cent of accessions in the pre-1980 period. At a 1981 IBPGR/FAO/UNEP conference, Latin American countries pressurised the IBPGR, successfully, to increase collections of crops less prominent in international trade.

By the end of 1981, despite opposition from US and UK representatives, a Resolution tabled by the Mexican delegation had been passed at the FAO calling for the FAO Director General to draft elements of an international convention on plant genetic resources, and investigate the feasibility of establishing a new international gene bank. Two years later, at the FAO's biennial conference, Jose Ramon Lopez Portillo, son of the former Mexican President, forced another vote which led to an International Undertaking on Plant Genetic Resources (IUPGR) and the creation of an International Commission on Plant Genetic Resources (ICPGR). The aim of these moves was transfer of control over genetic resources from the developed countries, and IBPGR in particular, to the United Nations.

Under the IUPGR, the notion of free exchange was to be respected, and it was not just landraces that were to be 'available without restriction', but also 'special genetic stocks (including elite and current breeders' lines and mutants)'. This angered the American Seed Trade Association (ASTA), who were represented in the US FAO delegation. They charged that the IUPGR struck:

> at the heart of free enterprise and intellectual property rights ... The definition includes unimproved and obsolete varieties, land races, wild and weedy species, all of which the seed industry believes appropriate to be preserved and freely exchanged. However, it also includes improved elite varieties and breeding lines within the definition of plant genetic resources ... This puts the Undertaking in direct conflict with the rights of holders of private property ... The anti-private business bias of the Undertaking is clear.

The IUPGR also proposed establishing a network of base collections under the jurisdiction of the FAO. Yet the Undertaking was a mild and voluntary agreement rather than a legally binding convention.

The Birth of Biotechnology

In the second half of the twentieth century, enormous strides have been made in the life sciences, particularly in the discipline which has come to be known as molecular biology. As a result of this growing body of knowledge, new commercial opportunities appeared on the horizon based on the use of tools developed through new discoveries in this field. The idea that plants could be made resistant to pesticides was no new idea. Wiebe and Hayes discussed it decades ago with regard to the reaction of barley varieties to the application of DDT. Yet work in biotechnology brought such a strategy closer to hand, raising the possibility of breeding plants designed to tolerate applications of proprietary chemicals.

The use of plasmids, in 1973 by Cohen and Boyer, to mediate gene transfer made possible a new alchemy. In the immediate aftermath, biotechnologists in the US showed awareness of public unease regarding this new technology by proposing a moratorium on certain types of research. Since 1977, guidelines laid down by the National Institutes of Health (NIH) have been progressively relaxed. The desire to regulate the industry has dwindled as authorities were persuaded of the commercial significance of the new technologies.

Reduced regulation of the biotechnology industry began to be perceived, and not just in the United States, as a means through which a country could maintain or improve its position in emerging bioindustries. Field reported that 'industrial competitiveness appears to represent the central and overriding concern of national strategies'.

From a different perspective, the United Nations Centre on Transnational Corporations opined that comparative advantage and the international division of labour were increasingly being shaped by technological prowess. Increasingly, regulation shied away from determining which technologies should be allowed for use, and the imperative of allowing new technologies to develop began to shape which regulations were considered acceptable.

The emergence of biotechniques for technology generation makes it possible to circumvent the constraints imposed upon genetic recombination by species incompatability. I argue below that formal agricultural research is undergoing a transition from the BCM mode to a biotechniques-mechanisation-legislative (BML) mode. This is not to imply that traditional plant breeding and chemical inputs are about to disappear from view, either now or in the near future.

The role of breeders, where they do not disappear altogether, is likely to undergo a change such that their work complements that of the biotechnologists, whilst the fact that to date, herbicide tolerance in crops is the most widely tested trait to date testifies to the likelihood of continued use of chemicals into the future. Nevertheless, the genetic determinants of interactions between the plant and various chemical and biological inputs are likely to become the focus of innovation in crop (and livestock) agriculture. The pivotal institutional innovation in enabling such a strategy to become privately profitable is IPR legislation.

Patenting Genetic Materials

In 1976, the first of the new biotechnology companies, Genentech, was formed by Herbert Boyer and venture capitalist Robert Swanson. In 1980, Genentech placed a share offering on the New York Stock Exchange, the prices of which shot up from $35 to $89 per share in twenty minutes, a record rate of increase. This was due to the fact that three months earlier, General Electric had successfully challenged an earlier decision by the US Patent and Trademark Office (PTO) which had ruled that an oil-degrading micro-organism developed by their scientist, Ananda Chakrabarty, was not patentable subject matter. The new ruling held that whether or not an invention was alive or dead was irrelevant to patent law.

In the PTO's ruling on Chakrabarty, the legal principle of 'pre-emption' disqualified materials protectable under the PPA or the PVPA from patent protection. But this ruling was also overturned in the 1985 *Ex parte Hibberd* case, in which Hibberd was granted patents on the tissue culture, seed, and whole plant of a corn line selected from tissue culture. Breeders could now choose the form of protection most suitable to them, including utility patents. The gene was being commodified.

Anxious to preserve its lead in the biotechnological race, the United States has moved fastest in bringing institutions into line with industry's desires. Employing both bilateral and multilateral channels, it has sought to harmonise standards across nations in line with its own structures, thus opening the way for global marketing of proprietary products of biotechniques. In November of 1982, at a ministerial level meeting held at the GATT at US insistence, the US proposed that the new round debate issues never before considered in earlier GATT rounds.

One such issue was trade in counterfeit goods, such as 'fake' Rolex watches, but the scope of this particular area was widened at the behest of the US and others to include the issue of IPR. Raghavan notes that this was 'thanks mainly to the negligence of the disorganised Third World countries, most of whom thought that it did not affect them'. This would not have been so critical had it not been for the fact that what many countries saw as a preparatory discussion was subsequently proposed by the US and others as the agenda for a new round. The inclusion of many new issues was given justification through addition of the prefix 'trade-related'. Many countries hoped that by stifling their objections to the inclusion of new issues such as 'services' and IPR, they would be rewarded with concessions on 'old' issues, such as tariffs on tropical products (and escalating tariffs on processed products thereof), textiles, and a continuation of benefits under the Generalised System of Prefences (GSP). A compromise text was agreed at Punta del Este in Uruguay at the end of 1986 which included IPR. Even then, developing countries refused to negotiate on the subject before the mid-term review in Montreal in December 1988, where agreement was reached concerning the negotiating agenda.

The GATT agreement was finally signed at Marrakesh in April 1994 despite the fact that market access negotiations had not been verified, and with many developing country negotiators complaining that they had seen the texts only weeks before. Ratification in many countries was rushed through with little debate, and on 1 January 1995, the new World Trade Organisation (WTO) came into existence. This would co-exist with the GATT until the end of 1995. The final text of the TRIPs (Trade Related Intellectual Property Rights) agreement establishes new multilateral rules on IPR based on uniform minimum standards for their protection and enforcement, including their availability, use and scope.

As regards plant materials, the treaty does allow for exemptions on grounds of perceived environmental or public order impacts, yet at the same time, the treaty states that plant varieties shall be protected by patents 'or by an effective *sui generis* system or by any combination thereof'. Although developing countries and least developed countries are allowed, respectively, five and ten years to implement the agreement, the '*sui generis*' clause is to be reviewed four years after the date of entry into force of the WTO agreement.

The TRIPs agreement offers little encouragement to communities that might seek to protect innovations which are the property of, as it were, the collective. The agreement recognises IPR as private rights, and also requires products to be 'capable of industrial application'. There is no mention of communities and their rights. Before the GATT negotiations even began, the US had made its intentions in respect of IPR abundantly clear through applying pressure bilaterally. Mexico was targeted as early as the 1970s, but little progress was made until, in the mid-1980s, the US began to link the issue of IPR reform to expansion of GSP concessions. By the end of 1986, Mexico had adopted a revised Patent and Trademark Law, though as a result of the efforts of domestic lobbying, this was deemed inadequate by the US administration.

In 1984, the US Trade and Tariff Act had been revised, S.301 of Title III of which invested the prevailing Administration with coercive powers aimed at righting 'unfair' trading practices. In 1985, cases against Brazil and South Korea were initiated, the former concerning, *inter alia,* copyright issues regarding software, the latter concerning failure to protect intellectual property. The same issue led to talks with China, whilst India has also come under pressure to reform its IPR legislation in the past. In January 1987, Mexico was informed of President Reagan's intention to withdraw $200 million of GSP benefits unless the perceived inadequacies of its new legislation were corrected. In 1988, the Omnibus Trade and Competitiveness Act was passed in the US. This included sections which came to be known as Super 301 and Special 301 respectively, retrospectively strengthening the coercive powers vested in the administration by the Trade and Tariff Act's 1974 revision. Under Special 301, a procedure was set up whereby the US Trade Representative could identify and initiate proceedings against countries considered to be offering inadequate IPR protection.

Within this list of countries, a Priority Watch List of countries was to be specified annually, and Mexico was on that first list in May 1989. When President Salinas de Gortari began his programme of liberalisation, and plans for a North American Free Trade Agreement were materialising, Mexico's stance on the IPR issue altered quite radically, and in 1990, when Mexico introduced a proposal for a TRIPs agreement at the GATT, it slipped off the 301 lists. Mexico's patent law was revised in June 1991 to explicitly allow for the patenting of plant varieties. It specifically addresses innovations likely to arise from the deployment of biotechnology.

As Fowler has made clear, GATT and the growing concerns of the US over intellectual property issues generally, enabled transnational corporations involved in agricultural biotechnology to have their concerns *vis-à-vis* the patenting of life addressed in new fora. In the case of the GATT, it became possible for the issue to be bundled up not only with concerns over patents and trademarks as they related to mechanical innovations, but also, since this was a take-all-of-it-or-leave-all-of-it package, with fourteen other areas with which the Uruguay Round was concerned.

Significantly, Watkins notes, 'The major actors in [the TRIPs] exercise have been the US-based Intellectual Property Coalition – a grouping of 13 major companies, including IBM, DuPont and General Motors – *and European agro-chemical giants such as Unilever, Hoechst and Ciba Geigy*.' Thus, IPR was being simultaneously harmonised and extended across the globe.

Upov

Paralleling the moves to enhance intellectual property protection under the auspices of the GATT were moves on the part of UPOV to bring the Convention into line with developments elsewhere, and particularly with respect to biotechnology. PBRs' research exemption made them inadequate for protecting biotechnically engineered plants since they offered protection at the level of the whole plant when what was required was protection at the level of the gene. But by 1987, it was clear that UPOV would be strengthened. According to Fowler *et al.*, UPOV's members had been divided between small seed houses and the integrated genetics supply industry, the former fearing gene patenting, the latter favouring new initiatives in this respect. UPOV was revised in March 1991.

Note that the right of farmers to save seed from one harvest for planting in the next, what the American Seed Trade Association had referred to as the 'farmers' right' in hearings on the PVPA, had become known as the 'farmers' privilege' and was no longer secure. Section 15.2 of the new Convention allows, as an optional exception, seed saving 'subject to the safeguarding of the legitimate interests of the breeder', implying that royalties should be paid to breeders where seed is saved.

On the other hand, there is a compulsory exemption for breeding other varieties. However, the interests of the breeder are, in general, strengthened since protection applies to 'Essentially derived and certain other varieties' as defined under Article 14 (5) (b) and (c). Lesser expresses concern that, since the definition is unclear, this will lead to quasi-IPR being granted to a breeder over thousands of attributes of a variety which he/she did nothing to create. The 'essentially derived' clause would appear to apply to genetic insertion, giving the owner of PBR the right to demand royalties from innovations based on insertion of one or two genes into a plant over which the right is held.

Increasingly, developed countries are bringing their PBR legislation into line with the 1991 UPOV Convention. Furthermore, although the TRIPs '*sui generis*' clause appears to offer room for manoeuvre in designing IPR for plant varieties, most believe that this translates into UPOV type standards and nothing less. Increasingly today, companies can choose which combination of protection they prefer, although European patent law currently forbids patenting of plant varieties.

Locking in to uniformity

Reflecting the above-mentioned events, the Crucible Group reported: 'Those who reviewed patent law a few decades ago may not recognize it today'. Probably no formal agricultural research organisation in the world has not at least cogitated upon the changes considered above, not all of whose are implications are, as yet, clear. It is important to understand at least some of these, and to contemplate the relevance of the changes to the existing BCM mode of agriculture.

Biotechniques and environmental critiques of agriculture are reported to be bringing about a reshaping of the technological development of agriculture. This is jumping the gun slightly. There are still unresolved questions concerning, in particular, consumer acceptance of genetically engineered products, which are already having an impact on the industry's development. However, Sharp may be right to talk of the laying of a new set of 'ground rules', making it 'inconceivable that those developing new drugs, new herbicides or pesticides, or new plant species, should not, somewhere *en route*, make use of gene cloning and sequencing techniques.' The implications would be that research which did not require such techniques might fall by the wayside.

A number of authors have commented on the paradigm-like shift that biotechnologies could achieve. Much of this discussion considers the issue at the macroeconomic level, and takes the view that it will not be biotechnology alone that leads to a new mode of accumulation, but biotechnology, the development of new materials, and information technologies working synergistically to form a new techno-economic paradigm.

There is no doubting that there could well be some revolutionary changes about to occur in the way in which the agro-food system functions. Most interesting of all are potential developments in the food processing industry, where some authors have speculated as to the possible emergence of a 'generic biomass inputs sector' as a result of technologies which allow biological materials to be fractionated into component parts for the final manufacture of food products. The implications for commodity markets as they are currently understood could be far-reaching. Other potentially revolutionary techniques relate to so-called novel products, which will affect the ways in which agriculture interacts with other sectors of the economy. Yet, whilst certain techniques used to create new products are certainly emerging, there appears to be substantial continuity with the past with respect to:

- Increasing horizontal integration across agricultural inputs – breeding for responsiveness to inputs and to facilitate harvesting will give way to herbicide tolerant varieties.
- Deepening of vertical integration – breeding has facilitated mechanised harvesting and handling of the final product. Biotechniques are increasingly geared towards downstream aspects of food production, representing higher value-added, and greater opportunities for profit, in upstream sectors of the chain of value in food. Lamola speaks of end-use tailored, or identity-preserved varieties.
- The actors involved are, in many cases, one and the same as those who prospered through the BCM paradigm (erstwhile agrochemical and seed companies).
- Emerging products take their cue from their supposed ability either to replace, or alter the functioning of, elements of the BCM paradigm which have been heavily criticised in the past.

- A continuing lack of emphasis, in private sector breeding, on pest resistance – although biotechniques provide tools for reducing pesticide use, current trends seem likely to increase, rather than reduce their use. Where resistance breeding is undertaken, it is of the gene-for-gene, vertical resistance type.

In many respects, therefore, the goals remain rather similar to those in the BCM paradigm. In particular, the attractions of the new techniques are seen principally in terms of the increased control that can be exerted over the transformation of organisms through recombinant DNA techniques. Indeed, Richards speaks of biotechniques as heralding a 'second designer phase' for agriculture.

Whereas the Green Revolution focused on ideotypes for monocropping in controlled *physical* environments, the second phase seeks to shape 'econotypes' to meet the need of future *economic* environments. As with the BCM mode, an emphasis on control within the labouratory has tended to obscure and marginalise the significance ecological issues concerning the functioning of biotechnically engineered products in the field.

The most obvious break with the past is the ability of biotechniques to extend the genepool available to breeders and biotechnicians beyond the primary and secondary genepoools into the hitherto unexplored (because of species incompatibility) tertiary genepool. Less immediately obvious, are the changes which have already been wrought by biotechniques on our perception of the nature of life itself, and the significance of these for our perception of the nature of food and agriculture in the longer-term.

Paradoxically, therefore, we are witnessing changes which are simultaneously profound, and incremental. The mode of agricultural research is changing through use of powerful new techniques, but its roots remain in the BCM mode.

This is to be expected if one accepts that agriculture had become locked in to the BCM way of doing things. As Teece points out, one aspect of the locking in process is that firms tend to do best what they have done in the past.

If the emergence of biotechnology constituted a radically new paradigm, learning advantages accumulated over time by established FAROs would have lost much of their significance. However, the fact that biotechnology is very much a process technology has meant that much of the significance of learning, particularly in the downstream operations of private multinational corporations, has been retained. Furthermore, as I have suggested above, the techniques are deployed in pursuit of a familiar goal, that of the genetically uniform ideal plant.

Changing Modes

The transition that is occurring can be understood through appealing to the framework developed by Freeman and Perez. As noted above, the BCM mode has come under fire for a variety of reasons, principally those associated with food quality and the environment. If the limits to the expansion of this mode had not yet been reached, such expansion was clearly under threat. The world market for agrochemicals saw three years of decline in the years 1991–93 before recovering somewhat in 1994 and 1995.

Farm support schemes, in the European Union and elsewhere, have begun to shift away from price support, which led to elevated levels of use (relative to that which would prevail with prices at world market levels) of agrochemical inputs, and towards conservation, often rewarding farmers for using fewer inputs. For Freeman and Perez, the transition from one techno-economic paradigm to another, brought on by the onset of recessionary trends, is characterised by:

The increasing degree of mismatch between the techno-economic sub-system and the old socio-institutional framework. It shows the need for a full-scale re-accommodation of social behaviour and institutions to suit the requirements and the potential of a shift which has already taken place to a considerable extent in some areas of the techno-economic sphere. This re-accommodation occurs as a result of a process of political search, experimentation and adaptation.

Once the socio-institutional framework matches the techno-economic sub-system, investment moves forward and growth is restored. For the BML mode to flourish, its techno-economic sub-system requires an appropriately matching institutional framework, including:

- a sympathetic regulatory framework for undertaking relevant research, including risk assessment methodologies as applied to the release of the products of gene technologies, and food safety legislation regarding genetically engineered food;
- political, social and environmental acceptance of the technologies and their end-products, reflecting confidence in the regulatory framework; and
- appropriate IPR protection, the importance of which is confirmed by, amongst others, Thelwall and Clucas, Caulder, Lamola, and Duffey.

A lack of institutional change will delay any upswing, and indeed, resistance from consumers concerning issues of health and environmental risk has been strong. Consequently, products have been slow to reach the market. Yet, for reasons elabourated below, it is the issue of IPR legislation which has greatest bearing on the issue of biological diversity in use in agriculture, and thus, the environmental risks posed by new biotechnologies in terms of vulnerability and the continued use of pesticides.

Those in the vanguard of the BML mode have sought to project it as environmentally friendly. In doing this, they have stressed the biological, *ergo* natural, characteristics of the work they are undertaking. The semantics involved have been illuminating, at one and the same time suggesting radical new possibilities (the economic and environmental attractions) and on the other, in an attempt to downplay the risks associated with the products of biotechniques (and the need for regulation), suggesting continuity with the past. Critics of biotechniques turn the matter around completely.

Whilst not disputing the fact that there is money to be made, they argue that the new possibilities should be reflected in the need for new forms of regulation, whilst continuity is likely to be reflected in the continuation of environmental problems. Their criticisms relate mainly to:

- The uncertainties in *ex ante* risk assessment associated with release of genetically engineered organisms into the environment, not least the difficulties in extrapolating from small-scale trials to large-scale field use, and problems associated with trade in the products concerned.
- The nature of individual products and their possible environmental and health consequences
- The environmental consequences of possibly increased uniformity.
- The political economy of the research being undertaken (who is it done by, and for?).
- The impact in terms of research not undertaken (a point well made by Rachel Carson in 'Silent Spring'.

These issues are not unrelated. The nature of the organisation funding research will determine the degree to which a notional social welfare function is reflected in their activities. Private organisations need not be concerned with social welfare, or only insofar as it affects profits.

The key to deepening private sector involvement in agro-food biotechnical research has been the extension of IPRs to living organisms. Governments have welcomed private sector participation in research, and have tended either to move the focus of their research away from near-market, and towards more basic, research, and/or to seek to take advantage of the patent system themselves to make financial gains from ongoing research.

Future for Diversity

As mentioned above, one of the criticisms levelled at those who believe that biotechniques herald a new 'sustainable agriculture' is that existing problems of uniformity will be exacerbated. Can the transition from BCM to BML mode re-introduce diversity into a system based on uniformity? From a purely technical view, biotechniques' capacity to draw upon genetic material from the tertiary genepool would suggest that additional genetic variation might be introduced. Thus, Bassett argues that new varieties 'will simply coexist with the old varieties: diversity will have been increased, not decreased'.

But this approach has two major shortcomings. Firstly, the basic research and the application are inextricably linked, so much so that possible applications are driving the direction of basic research. Thus, a growing proportion of public sector research is supporting commercially oriented research. Secondly, what actually happens is a subset of what could happen, as the earlier case-studies have argued. Kloppenburg bluntly states: 'the baby of biotechnology is not so easily separated from the corporate bath water,' which is exactly why, as Lesser points out, trends towards uniformity predate the existence of PBR and patents.

Strengthening IPR as applied to living material has encouraged private industry to engage in biotechnical research at levels above those that would have existed in their absence. Furthermore, because Governments now see biotechnical prowess as important for maintaining competitiveness, public research is beginning to resemble privately sponsored research, either through its increasingly subservient position to private industry, or through more overt aims to generate revenue from patentable research outputs. Erstwhile President of Harvard University, Derek Bok, has expressed concern that Universities will increasingly 'differ from corporations only because there are no shareholders and no dividends'.

An element of historical contingency is at work here, since many governments are cutting state spending on education and research where it is perceived to be of low market value. To the extent that environmental issues are intimately related to issues of social welfare, and because many environmental costs are not captured in market transactions, one assumes that private organisations are, notwithstanding their own public relations, less likely to integrate environmental issues into their research programmes.

Indeed, one survey, aimed at eliciting the ranking of breeding companies' priorities, placed the environment at the bottom of seven criteria. However, the same is increasingly true of publicly funded research.

Institutional changes are making the public / private distinction irrelevant in predicting the social welfare goals which will be pursued by one or other form. This is one reason why, notwithstanding some of the claims made for the efficacy of biotechnologies (most of which, incidentally, are made with the BCM technologies as the implied baseline), IPR will if not increase, then maintain, the vulnerability of agriculture in the field. Other reasons include the following:

Distinctness of Varieties

Already, anecdotal evidence suggests that seed companies rely on a few elite cultivars in their research programmes and new varieties are developed through minor modifications to these. Very little hard evidence is available, but it is clear that the number of varieties available (i) does not reflect genetic diversity, and (ii) masks the concentration in varietal use out in the field.

The future strategy for breeders will be structured by the IPR legislation in place. The combination of UPOV 1991 and patents ensures that the work of the breeder, and the value of an identified gene, is both recognised. The 'essentially derived' clause was introduced to deal with the problem of genetic distancing. Biotechniques make it possible to reduce the genetic distancing required to discriminate between varieties, in which case, the genetic variation existing in the field would become even further divorced from consideration of the number of varieties grown.

For example, since distinctness could now be measured at the level of the gene, a superfluous (in agronomic terms) gene could be spliced into a variety thus making it, potentially, distinct. As Smith points out, the practice of reverse engineering of varieties is increasingly common and would make the aforementioned practice more likely, rendering the granting of PBR meaningless. The 'essentially derived' clause, though of theoretical value, raises important questions of definition. For both Espenhain and Smith, who, in his excellent account, notes that numerous controversies in this regard are in no one's interest, case law will provide the answers.

Because UPOV 1991 extends the breeder's right to the commercialisation of essentially derived varieties (the principle of dependence), companies using genetic transformation techniques will either work with their own PBR-protected varieties, or license genes of interest to other companies for incorporation into their varieties. Strategic alliances between those specialising in biotechniques and those with greater specialism in traditional breeding seem likely.

Hence, Pioneer markets both soyabeans containing Monsanto's Roundup Ready gene (tolerant of Monsanto's glyphosate herbicide) and soyabeans containing the DuPont-owned sulfonylurea tolerant gene. DEKALB also has cross-licensing agreements with both Monsanto and DuPont. Those licensing technologies will seek to gain from the technology premium which biotechnically developed products aim to attract.

To a significant degree, varietal make-up will remain as before, but with genes spliced into a particular variety's background, and with plants themselves being made more uniform. However, since industrial structure will be affected by the evolving IPR framework, as well as the techniques themselves, and developments in individual sectors, there may be implications for the diversity of what is offered to farmers. This is considered in what follows.

Research Concentration

Patent enforcement prevents companies from carrying out research on a patented genetic sequence or process without paying royalties to the patent holder, whilst essentially derived varieties are subject to PBR under UPOV 1991. The strategic importance of patents for any company depends to a considerable degree on the extent of exclusion implied by the text of the patent. In this context, the current trend towards granting broad patents to companies is worrying indeed.

For example, Agracetus, a subsidiary of W. R. Grace (and recently taken over by Monsanto), was granted patent rights in the US over any genetically manipulated variety of cotton, and by the European Patent Office over all genetically transformed soybeans (with those for rice, groundnut and maize pending). Agrigenetics' patent on high oleic acid sunflowers effectively stopped all such work in this area outside the company. The prevalence of 'driftnet patenting' is at odds with the view held by many that patents encourage innovation. It raises the possibility that the seed industry for any one crop may ultimately become dominated, or at least hostage to, one commercial enterprise.

Patent enforcement is an important tool in building corporate empires and eliminating competition. Monsanto, which has staked much on its quest to become the 'Microsoft of engineered foods', has acquired companies, and stakes in others, reflecting its belief that patents will be a key source of competitive advantage in coming years. Pioneer's moves to patent its in-bred lines on grounds that this would prevent other companies carrying out research on them illustrates that even the 'biological patent' which hybrid corn varieties are endowed with is being superseded by strengthened IPR legislation. As patents proliferate, it will become increasingly difficult for any enterprise to conduct research in the full knowledge that it is legitimate, especially since burden-of-proof legislation makes it incumbent upon those accused of patent violation to prove that they are innocent. The possibility arises in which a company carrying out research happens, by accident, to be working with a variety in which a patented gene sequence exists. Such a company would be unwittingly breaking the law, and would be expected to provide proof of its innocence.

There is a suggestion that IPR-related concerns are driving strategic alliances in the industry. Since, increasingly, more than one form of IPR will be involved in developing a given variety, such alliances reduce the likelihood of any one IPR-holder blocking development of the product concerned. IPR-based restructuring can be expected to produce 'many casualties, some survivors, and a few successes'. The growing significance of IPR, appears to be leading to greater concentration in breeding effort, which is unlikely to promote diversity in agriculture.

Loss of Farmers' Privilege

UPOV 1991 appears to deny that farmers might also be breeders. Industry estimates suggest farmer-saved seed accounts for between one and two thirds of all seed planted in the world, though the proportion for developing countries is believed to be of the order 85 per cent. Indeed, in developing countries, the exchange of seed from farmer to farmer is probably the main avenue for diffusion of new seed varieties.

Erstwhile Director General of GATT, Peter Sutherland, has suggested that such informal practices are 'generally not of interest to the owners of protected varieties', yet already in the US, court actions have been taken against farmers involved in such activity, and it will not have been lost on IPR owners that interfarm sales constitute 62 per cent of all seed purchases in India where Sutherland made his speech.

In the US, companies such as Monsanto require growers of their Roundup Ready Soybeans to be licensed to grow the material, though other companies, including AgrEvo and DuPont, do not require licensing since, in the words of one commentator, 'they seem to feel that the additional chemical usage that's tied in with [DuPont's] STS beans is enough'. Monsanto states that 'if necessary, the terms of the [Roundup Ready] contract will be enforced under the P.V.P.A., US Patent Law and general contract law'. One study estimates that within a few years, 40 per cent of all US farmers will be contract growers, or renters of germplasm from the same companies to whom they sell their product.

Farmers will also need to be alive to the possibility, especially when growing out breeding crops (those which cross-pollinate) of falling foul of patent legislation. It is not clear, given reverse burden of proof, how the law would interpret a situation where a farmer grew a variety which through cross-pollination contained a patented sequence. Potentially, the onus will be placed upon farmers to ensure no such cross-pollination takes place. Furthermore, the possibilities for farmers to experiment with varieties covered by UPOV 1991 seem limited.

As with the case of 'essentially derived' varieties, it seems likely that case law will determine what is and is not allowed, but in the meantime, farmer experimentation with new varieties, which can create new races of out-breeding crops, may be a risky enterprise. This may have implications for diversity. More recently, the patent awarded to Delta and Pine Land Company on so-called 'terminator technology' (which prevents seeds from germinating in the next generation) provides a technological means through which to prevent farmer seed-saving.

Wide Use of Agronomic Genes

It has been stated that the introduction of novel genes will increase genetic diversity in agriculture. However, if the same gene is licensed to several companies for use in a large proportion of varieties in use, and in different crops, the gene becomes a component of uniformity. The possibility arises of the occurrence of a southern corn leaf blight on a more global scale. In China, 15 million hectares are planted to hybrid rice, each plant possessing, as with US corn varieties in 1970, a common gene for cytoplasmic male sterility.

Genes from the bacterium, *Bacillus thuringiensis*, which produce a protein that is lethal to some insects upon ingestion, are of great interest to corporations involved in biopesticides. Yet there are already concerns for insects' resistance to a number of strains of the bacterium. Herbicide tolerant genes could quite conceivably be transferred into vast areas of crop land, potentially increasing the vulnerability of crops globally, and leading to heightened problems with herbicide resistant weeds. Indeed, the relay race mentality of the BCM mode is accepted as a matter of course: 'in 50 years, biotechnologists will almost certainly still be developing new batteries of pest- and disease-resistant genes'.

The aim to engineer plants with tolerance to herbicides is a goal of companies that have integrated crop protection and seed production. In this way, purchasers of seed would be locked-in to the purchase of proprietary chemicals. This strategy reflects the fact that the costs of developing new agrochemicals are increasing owing to costly approvals processes. The costs of engineering the seed to lengthen the effective life of a given chemical compound are less, whilst it is also possible to extend the patent life of chemicals coming 'off patent' by specifying use of the proprietary form of a generic compound. The industry claims to have shifted emphasis to compounds of lower toxicity, although much attention has focused on glyphosate, which was listed by the US National Academy of Sciences as a potential carcinogen in 1987.

Ecological Interactions

One of the major fears of environmentalists is that a gene which has been transferred to a variety to enhance its competitive performance may be transferred sexually into wild relatives, especially where maize, potatoes, rice, chickpeas and common beans are concerned.

In these crops a wild-weed-crop complex is observed in which there is continual gene flow between wild and cultivated forms. More speculatively (the processes are poorly understood), horizontal gene flow mediated by micro-organisms may occur. In either case, the transfer of one gene may be sufficient for a plant to become invasive.

Lack of Research on Diversity

Quite apart from the tendencies remarked upon above which might exacerbate trends towards uniformity, the simple fact remains that little if any research associated with the use of biotechniques is being undertaken to encourage the use of diversity in the field. Indeed, biotechniques make it possible to clone plants and seeds so eliminating what residual variability there may have been in a crop bred using traditional methods.

Existing Seed Marketing Legislation

Perhaps most importantly, and what may in time become the most concrete expression of the way in which biotechniques affect the offerings of the seed industry to farmers in the fact that over time, varieties which are not the product of genetic manipulation will slip off existing National Lists of seed varieties. Since those on the list are the only ones which can enter into commerce, slowly it will become impossible to purchase seeds which are not genetically engineered.

This will force farmers and consumers alike, irrespective of concerns regarding genetic engineering, to purchase genetically engineered seeds. To the extent that genetic engineering concentrates on the integration of specific sequences within existing elite cultivars, irrespective of the number of cultivars made available to farmers, the genetic diversity within farming may decline, and certainly seems unlikely to increase.

Possibly, IPR will not increase uniformity. One scenario would see patented genes integrated into the background of existing varieties, and no change other than those created by the introduced genes. But this scenario assumes an unlikely scenario in which changes wrought by IPR and biotechniques will leave the seed industry unchanged in other respects, a scenario which current trends suggest is unlikely. Our inability to measure diversity, and the fact that we do not know where we stand today, makes it impossible to assess change on the basis of any reliable baseline. However, it seems reasonable to suggest that biotechniques are taking agriculture in different directions to those which would be implied by the alternative approaches.

If the BCM mode was environmentally damaging, and if there remain unanswered questions regarding the impact of the BML mode, why does this new mode appear to be gaining the support of most FAROs, public and private, especially when alternative paths exist?

Alternative agriculture

In Mellon's words 'the hype surrounding biotechnology diverts our attention from those [pesticide free] solutions by focusing attention on technologically dazzling new products. By setting the proper goal, we will avoid the danger of spending millions trying to genetically engineer ten "better" pesticides, when for far less we could have taken our agriculture systems off the pesticide treadmill forever.'

In particular, the alternatives (re-)emerged with some force in the 1980s on the back of environmental concern seem likely to remain alternatives in terms of the resources devoted to them. The transition from BCM to BML mode must be understood in this context. This was not a transition that was inevitable. The BCM was under fire, and alternative approaches beyond biotechniques were available.

If IPR was essential to the BML mode, they may be decidedly unhelpful for those seeking to do research outside that mode. Sederoff and Meagher opine that IPR 'are having a dramatic negative effect on the progress of non-profit research'. The following sections consider debates concerning genetic resources which have taken place outside GATT and UPOV.

Recognition for farmers' rights

The aim to resist IPR strengthening has been closely related to attempts to encourage the use of diversity in agriculture. Whilst the agricultural biotechnology companies had their sights on institutional changes allowing for the granting of IPR over life through the GATT, the FAO was debating the issue of the rights of farmers over germplasm, especially from 1987 onward.

Farmers' Rights would be the counterbalance to the spread of PBR. In 1987, an International Fund for Plant Genetic Resources was set up, and was legally established in 1989. The Fund was designed for genetic conservation and utilisation work, and administered by the ICPGR. Mexico argued that donations to such a Fund should be mandatory in the same way as are royalties to a patent holder, but no agreement was forthcoming. By 1989, developing country governments had let it be known through negotiations at the FAO that failure on the part of developed country governments to acknowledge the concept of Farmers' Rights would result in those countries being denied access to developing country genetic resources. Similarly, developed country patents would not be honoured.

In what was effectively an exercise in horse-trading, at the 1989 meeting, developed countries insisted on an additional Resolution modifying the International Undertaking such that it recognised PBR. 'Free accesses to germplasm explicitly would not mean 'free of charge'. In return, a Resolution on Farmers' Rights was passed recognising the rights of farmers in respect of their work conserving, improving, and making available genetic resources.

In the midst of an increasingly polarised debate, a notable event was the Keystone Dialogue held at the Keystone Centre in Colorado in 1988, at Madras in 1990, and Oslo in 1991. Major transnational corporations, NGOs, IBPGR, national genetic resource programmes, the academic community, and the Rockefeller Foundation reached the following consensus conclusions after the second meeting:

- IPR were credited with encouraging development of new varieties, but also with encouraging genetic uniformity and erosion;
- Attempts to include IPR for plants under the auspices of the GATT negotiations were criticised; and
- Recognition of Farmers' Rights, and commendation of the idea of a Fund such as that extant at the FAO as a means of providing a form of concrete recognition thereof.

Corporate delegates refused to sanction a compensation mechanism, merely a fund recognising Farmers' Rights, a position which some participants would have found unacceptable were it not for the fact that it was agreed the fund should be mandatory. Furthermore, rather than the figure of $150 million proposed by NGOs at the FAO, the consensus figure arrived at was $500 million.

In November 1991, another Resolution concerning genetic resources was passed at the FAO. This amendment to the Undertaking upheld 'that nations have sovereign rights over their plant genetic resources and that breeders' lines and farmers' breeding material should only be available at the discretion of their developers during the period of development'. Although Farmers' Rights were recognised, they were given no substance. The International Fund, legally established in 1989, failed to materialise. Only in India have attempts been made to give substance to the concept through taxation of seed industry income. Farmers' Rights still amount to little more than a polite thank you to farmers who have conserved genetic diversity *in situ*. This was most clearly illustrated in discussions at the June 1996 Leipzig Conference where the issue of Farmers' Rights showed that developed country donors were reluctant to support *in situ* conservation, partly, one suspects, because of issues related to sovereignty in respect of genetic resources (see next section).

The Convention on Biological Diversity

National sovereignty over genetic resources was a feature of the Convention on Biological Diversity (CBD), which entered into force as a legally binding international treaty at the end of 1993. The CBD is a framework convention whose objectives, stated in Article 1, are: the conservation of biological diversity, the sustainable use of its components and *the fair and equitable sharing of the benefits arising out of the utilization of genetic resources, including by appropriate access to genetic resources and by appropriate transfer of relevant technologies,* taking into account all rights over those resources and to technologies, and by appropriate funding.

Article 3 of the CBD lays down the principle that:

States have, in accordance with the Charter of the United Nations and the principle of international law, the *sovereign right to exploit their own resources* pursuant to their own environmental policies, and the responsibility to ensure that activities within their own jurisdiction or control do not cause damage to the environment of other states or of areas beyond the limits of national jurisdiction. This effectively confirms that the free exchange principle is something of the past.

For corporations, sovereignty appeared to cede too much control to governments over genetic resources which were of increasing value to biotechnology companies. For non-government organisations, this debate seemed to miss the point that it was not states who were really responsible for maintaining biological diversity within their borders, but local communities. The CBD offers little for local communities, and does not explicitly recognise Farmers' Rights, though the role of indigenous communities in conserving biodiversity is recognised in the preamble.

Articles 12, 17, 18 and 19 each refer to aspects of the biotechnology debate, but Article 16 of the CBD, dealing with transfer of technology, was the most heavily negotiated. Essentially, the debate centred around the fact that developing countries would most likely be providing raw materials for a biotechnology industry seeking to patent innovations. Article 16.2, suggests that:

In the case of technology subject to patents and other intellectual property rights, such access and transfer shall be provided on terms which recognise and are consistent with the adequate and effective protection of intellectual property rights. The application of this paragraph shall be consistent with paragraphs 3, 4 and 5 below.

This Article was essentially a compromise meant to defuse the situation as regards the way in which technology transfer should account for IPR. These latter paragraphs make provision for the transfer of technology to developing countries on 'mutually agreed terms'. Article 16.5 suggests that the Contracting parties: recognizing that patents and other intellectual property rights may have an influence on then implementation of this Convention, shall cooperate in this regard subject to national legislation and international law in order to ensure that such rights are supportive of and do not run counter to its objectives.

The CBD's equivocation on IPR issues was perhaps the main reason why George Bush's US delegation felt unable to support the CBD in Rio. Ever since the late 1980s, when methods of screening increased in sophistication and fell in cost, plant research has acquired new significance for pharmaceutical companies. Thus, head of the US delegation, William Reilly, stated: 'We have negotiated in the Uruguay Round of GATT to try to protect intellectual property rights. We're not about to trade away here in an environmental treaty what we worked so hard to protect there'.

The CBD does not apply retrospectively. Thus, the legal status of *ex-situ* collections of resources donated by developing countries but housed in the developed world was left unclear. In 1994, the World Bank appears to have sought to prevent the CGIAR's *ex situ* collections from falling under intergovernmental control by taking control of these itself in exchange for new funds for the CGIAR.

However, 112 governments unanimously, and ultimately successfully, called for establishment of intergovernmental control over the CGIAR *ex situ* collections. The status of other collections held in developed country genebanks was discussed at the FAO's Leipzig Conference, but the outcome gave little encouragement to those countries which donated germplasm in the first place, as exemplified by the attempts of the company PHYTERA to gain access to genetic resources collected in developed country botanical gardens.

Indigenous Peoples

As so often in the past, indigenous peoples have been forgotten in the bulk of negotiations which affect a resource maintained largely by them. Their concerns span both the prime focus of the CBD and the issue of Farmers' Rights, as well as the TRIPs negotiations. Increasingly, the contributions made by indigenous peoples in terms not just of germplasm, but also knowledge regarding its use, are recognised. Yet this recognition has led to little concrete action to protect their interests. In response principally to the heated debate generated by issues related to the patenting of indigenous people's cell lines, the Indigenous Peoples Biodiversity

Network was formed with the objective of safeguarding their interests with respect to biodiversity and their knowledge thereof. Out of the debate concerning Farmers' Rights and IPR as they affect indigenous peoples has come awareness that current formal systems do not adequately recognise indigenous knowledge systems, in which knowledge is often held at the level of the community. More and more companies are screening plant materials for useful products, yet it is estimated that the chances of finding useful products are at least doubled if indigenous, or folk knowledge is utilised. An authority on the issue of IPR and indigenous peoples, Darrell Posey, notes that IPR pose seven problems for indigenous peoples:

- They do not grant rights to collective entities;
- They protect unique acts of discovery rather than transgenerational knowledge (from, for example, spirits, vision quests, or oral transmission) which tends to be public;
- They do not recognise non-western systems of ownership, access and tenure;
- They aim to promote commercialisation whereas the aims of indigenous peoples may be to prevent such activity;
- They recognise market values only and not spiritual, aesthetic, or cultural value;
- They are, as is clear from the above, intimately bound up with power relations;
- They are expensive to obtain and difficult to defend.

Posey goes on to cite a number of examples where indigenous peoples have displayed almost uniform hostility to what they perceive as an insidious trend in IPR legislation which seeks to deepen the exploitation of their resources and their knowledge. They perceive simple recognition of their contributions, as exemplified by the Farmers' Rights issue, as typically patronising in the face of a continued absence of legal mechanisms adapted to meet their needs and concerns.

RAFI explores a number of ways in which indigenous communities could find space within existing legislation to protect their innovations. They argue that: There is a strong case to be made that the uncompensated appropriation of farmers' varieties and medicinal plants constitutes real theft and that the parties responsible should be pursued under criminal law at the expense of national law enforcement agencies in the country where the theft occurs (the patenting country). In the face of the forces mentioned above, it seems unlikely that much room for manoeuvre exists for those who would seek to place Farmers' Rights on the same level as IPR. Some countries have, however, been exploring the potential for exploiting the '*sui generis*' clause mentioned above in designing alternative IPR regimes which allow, for example, for communities, and not just individuals, to make IPR claims. As pointed out by Allen, much innovation is the product of collective rather than individual efforts. In India, the concept of Collective Intellectual Property Rights is gaining credibility, whilst the Andean Pact is committed to developing a regime on collective rights of indigenous peoples.

This chapter has sought to show, through examining evolving IPR regimes, that a new mode of development of agricultural technologies is emerging. Some, indeed most of its most vocal and powerful supporters are drawn from the leaders of the BCM mode. The commercial possibilities presented by biotechniques saw pressure to extend IPR schemes. Following the 1980s, a decade which saw recognition gained in an international forum for the concept of Farmers' Rights, the rights of farmers to save seed first underwent conversion to a 'privilege', and were then consigned to history.

The old BCM mode still prevails despite the attacks of environmentalists and the growing awareness of available alternatives. If the possibilities for alternatives to thrive alongside the BCM mode seem limited, they are likely to be more so as the BCM mode is superseded by BML techniques. Although the BML mode's supporters have often appeared as keen as environmentalists to see the back of the old BCM mode, their agenda is not an alternative based on bringing diversity back into the picture, but the ushering in of new techniques aimed at increasing the potential for achieving a genetically uniform ideal plant.

Major suppliers of agrochemicals condemn the technologies for which they themselves have been responsible as manifestly unsustainable. In answer to his own question whether such companies are 'Planetary patriots or sophisticated scoundrels?' Kloppenburg writes: 'Having been recognised as wolves, the industrial semioticians (and you thought they were only manipulating genes!) are now redefining themselves as sheep, and green sheep at that.'

6

Agriculture Biotechnology in 21st Century

Food security

In a world where population growth is outstripping food supply agricultural- and especially plant-biotechnology, needs to be swiftly implemented in all walks of life. The world population is expected to reach 7 billion within 25 years, over 10 billion in the year 2050, while agricultural production is growing at the slower rate of about 1.8 percentage annually. All human beings depend on agriculture that produces food of the appropriate quality at the required quantities.

Since this can no longer be achieved by traditional methods alone, *breeding through plant biotechnology is a necessity*. In the long run the massive and immediate implementation of plant and agricultural biotechnology is more highly crucial than that of medical biotechnology, since more people worldwide die from famine and diseases related to malnutrition than from "modern", western diseases. With "Disaster Management" becoming a central issue in modern sociology research and curricula, *food and food product shortages are the ultimate disaster*. However, unlike most natural disasters, this is one that we can prepare for and even prevent.

Domestication of plants and animals found in the wild, combined with gradual long-term changes in their qualitative and quantitative traits, were the first attributes of agriculture. Domestication, followed by food storage, coincided with the growth of microorganisms. Thus was born classical food fermentation, the earliest known application of biotechnology for the generation of food products. This traditional agriculture now faces several serious limitations:

1. Market limitations: The world is becoming a global village whose free-market rules negate the effectiveness of local pricing policies, and where a dictate of international trade and policies exists. This has greatly affected future developments in agriculture, still the world's largest business.
2. Limitations of natural resources: Global climatic changes (resulting mainly in desertification and salinization), industrialisation and urbanisation, have reduced land and water availability and caused alarming deterioration of soil, water and air quality.

3. Inherent biological genetic limitations: Although previously highly efficient, the release of new improved genotypes by classical breeding is now too slow to cope with the demands, and is considerably limited by the lack of appropriate "natural" genes that can be introgressed by traditional genetic crosses.

Only two major potential solutions seems to exist for increasing food supply and agricultural commodities, in addition to continuously improving agricultural practices, despite the aforementioned limitations:

1. A search for alternative food sources (e.g., marine or extraterrestrial products),
2. Enhanced efficient plant breeding.

Combining biotechnology with classical physiology and breeding

Food production, for both quantity and quality, as well as for new plant commodities and products, in developed and developing countries around the globe, cannot rely solely on classical agriculture. Human survival, vis-à-vis a continuous increase in agricultural productivity, depends on the effective *merging of classical breeding with modern plant biotechnology* and the novel tools it provides. The "green revolution", for example, increased wheat production 10-fold in India and several other countries in South East Asia, thereby feeding three times as many people. However, this revolution has already been exploited to its limits, and alternative solutions are required to breed improved crops.

Now biotechnology, integrated with classical breeding, is on the verge of creating the "evergreen revolution". The potential to improve plant and animal productivity and their proper use in agriculture relies largely on newly developed *DNA biotechnology and molecular markers.*

These techniques enable the selection of successful genotypes, better isolation and cloning of favourable traits, and the creation of transgenic organisms of importance to agriculture. Together, these generic techniques are both an extension and an integral part of classical breeding, contributing successfully to shortening breeding and selection cycles.

The new plant biotechnology implies the use of recombinant DNA techniques and in vitro cell biology in three major areas:

1. As an aid to classical breeding: This includes the ongoing genome mapping projects, e.g., in Arabidopsis, rice, maize and tomato, combined with the recent activities in functional genomics, proteomics and bioinformatics, and DNA marker-assisted selection. The combined use of these techniques will soon shorten the time required for "classical" breeding and selection cycles.

2. Generation of engineered (transgenic) organisms: In view of the inherent limitations of introgressing new genes by traditional genetic crosses (i.e., lack of appropriate desired genes and crossing barriers), the efficient engineering of plants has already resulted in improved field-grown transgenic plants in several important crops. The result of this impressive development, which began only 18 years ago, have made possible the direct insertion and integration of genes isolated from several organisms, and the creation of novel, and otherwise impossible genetic recombination.
3. Integration of microorganisms into plant production systems: The biotechnological development of new symbiotic, antibiotic, and antagonistic relationships between plants and microorganisms (fungi, bacteria and insects) using, among other techniques, engineered plants and microorganisms, creates new possibilities. Some of these include biological control of pests, biofertilization and plant growth stimulation, and bio- and phytoremediation.

During the last two decades, these new biotechnologies have been adapted to agricultural practices and have opened *new vistas for plant utilisation*. This will continue and intensify in the next decade. Plant biotechnology — especially in vitro regeneration and cell biology, DNA manipulation and genetic modification of biochemical pathways — is changing the plant scene in three major areas:

1. Growth and development control (vegetative, generative and propagation),
2. Protecting plants against the ever-increasing threats of abiotic and biotic stress,
3. Expanding the horizons by producing specialty foods, biochemicals and pharmaceuticals. These areas were extensively discussed at the 9th international congress of the IAPTC and B "Plant Biotechnology and In Vitro Biology in the 21st Century", held in Jerusalem in June 1998.

Vegetative, Generative and Propagation

The better insight into the control of plant regeneration, morphogenesis and patterns of cell division achieved during the last two decade, is due to three major discoveries:

1. The *totipotency and regeneration ability of plant cells and tissues*, as revealed by cell culture and micropropagation,
2. The elucidation of *genes responsible for hormone production and activation* in plants,

3. Active research into the mechanisms and molecular control of the *cell cycle and signal transduction pathways,* in part adopted from previous studies with animal cells, in part unique for plants. These have enabled both the *control and biotechnological manipulation* of vegetative growth, generative patterns (e.g., of flowers and seeds) and of micropropagation.

Vegetative Growth

Morphogenetic control mechanisms are still extremely obscure, but the advent of molecular hormone and cell-cycle research is sure to lead to a better understanding of vegetative growth patterns. Thus, the possibility of biotechnologically manipulating *plant growth rate and architecture* can become a reality. For example, potential consequences of controlled auxin overproduction/availability include: adventitious root formation of importance to propagation, cell and organ elongation for biomass production, increased apical dominance of importance to timber production, etc.

Controlled cytokinin overproduction/availability can result, among other things, in enhanced bud break — which is of great importance to plant architecture, branching and compactness — a desired characteristic for some ornamentals, and delayed leaf and plant senescence. No less important, in this respect, is the potential - as yet not practical - of affecting the orientation and rate of cell division, cell elongation and tissue longevity, by interfering with the cytoskeleton and cell cycle, the synthesis of cellulose and other cell components, and programmed cell death, respectively. A few of these possibilities have already been realised.

Generative Development

Flowers, fruits and seeds are extremely important for agriculture. Hence, biotechnological research and development aims to interfere with and control their development and characteristics, and some of the many related studies have already produced practical applications. The major targets in *flower development* are colour, scent and senescence.

Strategies for the molecular breeding of flower colour and scent include over- and underexpression of colour (anthocyanins and carotenoids) and scent (volatiles) compounds, with respect to their biosynthesis, cellular transport and targeting. Important targets for controlling *fruit development* include growth, ripening and senescence (as for vegetative growth), colour and scent (similar to flowers) and, in addition, flavour — particularly metabolic control of sugar, acid and other flavour components. Of great importance to fruits are biotechnological strategies for the production of seedless fruits via parthenocarpy (overproduction of auxin), pollen destruction (no fertilisation), or arrest of embryo development. The manipulation of *seed development* using biotechnological strategies is especially critical, since the seed industry (together with vegetative propagation material) constitutes the germplasm of the future for any type of plant production system.

Seeds and vegetative propagules are, practically speaking, packages of genes that form the basis of all advanced and economically viable agricultural industries, both national and private. Biotechniques and molecular strategies are now available for the major seed-based operations: hybrid seed production, generation of artificial seeds (coated somatic embryos), and for the establishment of germplasm banks that may solve some of the biodiversity issues.

Micropropagation

Micropropagation is used routinely to generate a large number of high-quality clonal agricultural plants, including ornamental and vegetable species, and in some cases also plantation crops, fruits and vegetable species. Micropropagation has significant advantages over traditional clonal propagation techniques.

These include the potential of combining rapid large-scale propagation of new genotypes, the use of small amounts of original germplasm (particularly at the early breeding and/or transformation stage, when only a few plants are available), and the generation of pathogen-free propagules.

This impressive application of the *principles of plant cell division and regeneration to practical plant propagation* is the result of continuous tedious studies in hundreds of labouratories worldwide, many of them in developing countries, on the standardisation of explant sources, media composition and physical state, environmental conditions and acclimatisation of in vitro plants.

Particularly noteworthy are the many recent studies on the molecular of organogenesis and somatic embryogenesis.However, further practical applications of micropropagation, which is also commercially viable, depends on *reducing the production costs* such that it can compete with seed production or traditional vegetative propagation methods (e.g., cuttings, tubers and bulbs, grafting).

Techniques that have the potential to further increase the efficiency of micropropagation, but still await further improvements, include: simplified large-scale bioreactors, cheaper automatization facilities, efficient somatic embryogenesis and synthetic seed production, greater utilisation of the autotrophic growth potential of cultures, and good repeatability and quality assurance of the micropropagated plants.

Abiotic and Biotic Stress Tolerance

The application of molecular genetics and plant transformation to the *diagnosis and control of plant pests* has become one of the major practical success stories of plant biotechnology in the past decade. The availability of dozens of transgenic crop plants which are resistant to a range of insects, viruses and herbicides, as well as to several phytopathogenic fungi and nematodes has been validated under both field and laboratory conditions, and is of great economic importance.

Moreover, applying the principles of engineering plants for resistance to these pests to other plants of agricultural importance is now considered routine, although in practice still labourious, especially for new genotypes. Apart from a wider application to additional plants, the real *challenges lying ahead* include:

1. Improved expression of the target genes in the plants, especially their spatial and temporal control,
2. The use of wide-spectrum and alternative target genes to circumvent the problem of pest resistance,
3. Intensified integration of biological control via the use of selected and engineered microorganisms with a biocontrol potential.

While plant biotechnology has been applied successfully to fighting a large number of pests, this is not yet the case for abiotic stress conditions such as drought, salinity, extreme temperatures, chemical toxicity and oxidative stress. *Drought and salinization are the most common natural causes of lack of food and famine* in arid and semiarid regions, and the most serious environmental threats to agriculture in many parts of the world.

Desertification, resulting from overexploitation by the local inhabitants, is often aggravated by regional climatic changes, and results in increased soil erosion and a decrease in land and agricultural productivity. It is estimated that increased salinization of arable land will have devastating global effects, resulting in 30 percentage land loss within the next 25 years, and up to 50 percentage in the year 2050.

Although more difficult to control and engineer than the usually monogenic traits of resistance to biotic pests and herbicides, the genetically complex response to abiotic stress is globally and regionally far more important. Therefore, *breeding for plant tolerance to drought and salinity stress* should be given a high research priority in all future agbiotech programmes. Strategies for the manipulation of *osmotic stress tolerance* in plants might include: expression of osmoprotectants and compatible solutes, ion and water transport and channels, expression of water-binding and membrane-associated dehydrins and other proteins, transcription factors and DNA-binding proteins, etc.

Also of specific interest are the intervening stages of stress perception, signal transduction (ABA and others), and protein modification. The discovery of new stress-related genes and the design of stress-specific promoters are equally important.

Food, Biochemicals and Pharmaceuticals

Traditionally, agriculture was targeted to improving the production of plant-derived food, in terms of both quantity and quality. This was also the initial primary target of plant biotechnology. The second phase of plant biotechnology is now gradually being implemented: a shift from the production of low-priced food and bulk commodities to high-priced, specialised plant-derived products.

This includes two major categories of biomaterials:

1. Direct improvement and modification of specialised constituents of plant origin,
2. The manufacture in plants of non-plant compounds. Biotechniques, mostly based on the *engineering of metabolic pathways*, are now available for modifying many plant constituents that are *used in the food, chemical and energy industries*. This includes many "primary" metabolites: carbohydrates (starch synthesis, yield and allocation, production of high-amylose or high-amylopectin starch, increased sucrose synthesis for the sugar industry, fructan production, etc.), proteins (improvement of amino acid composition and protein content), oils and fats (ratio of saturated to nonsaturated fatty acids, increased content of specific valuable fatty acids like erucic acid, ricinoleic acid and others).

Many other plant constituents are either minor or non-food components, but have specific high- value applications, such as specific fatty acids as an alternative energy source, polysaccharides with heat hysteresis properties and for bioaffinity purification, temperature and salt-resistant enzymes for the food industry, etc. Moreover, the use of plants as *"bioreactors" for the production of "foreign", non-plant compounds* is gaining momentum and may eventually lead to alternative types of agriculture.

This includes, for example, production of bioactive peptides, vaccines, antibodies and a range of enzymes — mostly for the pharmaceutical industry. For the chemical industry, plants can be used to produce, e.g., polyhydroxybutyrate for the production of biodegradable thermoplastics, and cyclodextrins, which form inclusion complexes with hydrophobic substances. Achievements today in plant biotechnology have already surpassed all previous expectations, and the future is even more promising. The full realisation and impact of the aforementioned developments depend, however, not only on continued successful and innovative research and development activities, but also on a favourable regulatory climate and public acceptance. About 12 percentage of the world's land surface is used to grow crops, and the agricultural area required to support food production — 0.44 ha / capita in 1961 — will probably have been reduced to 0.15 ha / capita in 2050.

The intensification of agriculture with its aforementioned limitations thus requires enhanced and more efficient plant breeding and the release of economical, high-return and patentable plant-derived products. This cannot be achieved without supporting advanced research and development in biochemistry, physiology, genomics and biotechnology of agricultural plants. Plant scientists now have a central role in society, not unlike their place 300 years ago when Jonathan Swift stated: "Whoever could make two ears of corn or two blades of grass to grow upon a spot of ground where only one grew before, would deserve better of mankind, and do more essential service to his country, than the whole race of politicians put together."

7

Importance of Agriculture Biotechnology

Genetically Modified Foods

It is our quest as technology education teachers to prepare our students to become technologically literate and productive members of our society—a daunting task in a society growing ever dependent on our technological development. The challenge becomes formidable in our attempt to not only understand the complexities and impacts these technologies might have on our society, and indeed on the whole human race, but at the same time impart an understanding to our students so that they have the opportunity to become technologically literate with the capability to make sound rational decisions in the use and deployment of these technologies for the betterment of our planet and all its inhabitants.

One area of technological development that cannot be ignored in our classrooms is that of genetically modified foods. The question is not whether we should modify the genetics of our crop plants, which has been going on since plants were first domesticated—the real questions should focus on how new genetically engineered modifications will affect our society and how genetically modified crops and plants will alter our environment. The term GM food is used in reference to crop plants modified for human and animal consumption using the latest molecular biological techniques. It is the purposeful addition of a foreign gene or genes to the genome of an organism.

A gene holds the information that will give the organism trait. Genetic engineering physically removes the DNA from one organism and transfers the gene(s) for one or a few traits to another. Since cross breeding between two sexually compatible plants is not necessary, the sexual barrier between species is overcome. These modern techniques now enable scientists to enhance desired traits such as increased resistance to herbicides or improved nutritional content, in ways they could not before, and with greater ease and precision.

Genetic engineering is not bound by the limitations of traditional plant breeding—not only can genes be transferred from one plant to another, but genes from non-plant organisms can also be used. The best-known example of this is the use of Bt genes in corn and other crops. Bacillus thuringiensis is a naturally occurring bacterium that produces crystal proteins that are lethal to insect larvae. These proteins have been transferred into corn and cotton so that these plants can create their own pesticides. Through genetic manipulation, crop plants are being modified for pest resistance, herbicide tolerance, disease resistance, cold tolerance, drought tolerance, salinity tolerance, and increased nutrition. Researchers are also working to develop edible vaccines in tomatoes and potatoes. Plants such as poplar trees have been genetically engineered to clean up heavy metal pollution from contaminated soil.

Genetically Modified (GM) foods are produced from genetically modified organisms (GMO) which have had their genome altered through genetic engineering techniques. The general principle of producing a GMO is to insert DNA that has been taken from another organism and modified in the laboratory into an organism's genome to produce both new and useful traits and phenotypes.

Typically this is done using DNA from certain types of bacteria. GM Foods have been available since the 1990s, with the principal ones being derived from plants; soybean, maize, canola and cotton seed oil. The first commercially grown genetically modified food crop was the Flavr Savr tomato which was made more resistant to rotting by Californian company Calgene.

Calgene was allowed to release it into the market in 1994 without any special labeling, where it was welcomed by consumers who purchased the fruit at two to five times the price of standard tomatoes.

However, production problems and competition from a conventionally bred, longer shelf-life variety prevented the product from becoming profitable.

A variant of the Flavr Savr was used by Zeneca to produce tomato paste which was sold in Europe during the summer of 1996. It's labeling and pricing were designed as a marketing experiment which proved that, at the time, European consumers would accept genetically engineered foods.

This attitude would be drastically changed after outbreaks of Mad Cow Disease weakened consumer trust in government regulators, and protesters rallied against the introduction of Monsanto's "Roundup-Ready" soybeans. The next GM crops included insect-protected cotton and herbicide-tolerant soybeans both of which were commercially released in 1996.

These crops have been widely adopted in the United States. They have also been extensively planted in several other countries (Argentina, Brazil, South Africa, India, and China) where agriculture is a major part of the total economy. Other GM crops include insect-protected maize and herbicide-tolerant maize, cotton, and rapeseed varieties.

Crops under development

The following GM crops are in development.

- Sweet potato resistant to the feathery mottle virus
- Further development of golden rice to increase levels or bioavailability of iron, zinc, vitamin A, vitamin E and improve the quality of proteins
- Maize with increased levels of the amino acid lysine and protein for animal feeds
- A variety of plants able to better tolerate non-biological stresses which are commonly encountered in a normal growing season, such as water and nitrogen limitation, or survive extreme growing conditions, such as high-salinity, drought, acidic soils, or hot weather. Such traits can provide more reliable crop performance over an extended period of cultivation.
- Transgenic rice has been developed by a Californian company to improve oral rehydration therapy for diarrhea. In sub-Saharan Africa and parts of Latin America and Asia, diarrhea is the second highest infectious killer of children under the age of five, accounting for some two million deaths a year. Recent 2005-6 trials in a Peruvian Hospital have demonstrated that specialized milk proteins lactoferrin and lysozyme made in transgenic rice plants improve the effectiveness of oral rehydration solution used to treat diarrhea.

Abundance of GM Crops

Between 1996 and 2005, the total surface area of land cultivated with GMOs had increased by a factor of 50, from 17,000 km² (4.2 million acres) to 900,000 km² (222 million acres), of which 55% were in the United States.

Although most GM crops are grown in North America, in recent years there has been rapid growth in the area sown in developing countries. For instance in 2005 the largest increase in crop area planted to GM crops (soybeans) was in Brazil (94,000 km² in 2005 versus 50,000 km² in 2004.) There has also been rapid and continuing expansion of GM cotton varieties in India since 2002. (Cotton is a major source of vegetable cooking oil and animal feed.) It is predicted that in 2006/7 32,000 km² of GM cotton will be harvested in India (up more than 100% from the previous season).

Indian national average cotton yields have been boosted to close 50% above the long term average yield during this period. The publicity given to transgenic trait Bt insect resistance has encouraged the adoption of better performing hybrid cotton varieties, and the Bt trait has substantially reduced losses to insect predation. Economic and environmental benefits of GM cotton in India to the individual farmer have been documented. In 2003, countries that grew 99% of the global transgenic crops were the United States (63%), Argentina (21%), Canada (6%), Brazil (4%), China (4%), and South Africa (1%). The Grocery Manufacturers of America estimate that 75% of all processed foods in the U.S. contain a GM ingredient. In particular, Bt corn, which produces the pesticide within the plant itself is widely grown, as are soybeans genetically designed to tolerate glyphosate herbicides. These constitute "input-traits" that financially benefit the producers, have indirect environmental benefits and marginal cost benefits to consumers.

In the US, by 2006 89% of the planted area of soybeans, 83% of cotton, and 61% maize was genetically modified varieties. Genetically modified soybeans carried herbicide tolerant traits only, but maize and cotton carried both herbicide tolerance and insect protection traits (the latter largely the *Bacillus thuringiensus* Bt insecticidal protein). In the period 2002 to 2006, there were significant increases in the area planted to Bt protected cotton and maize, and herbicide tolerant maize also increased in sown area.

Future Developments

Future envisaged applications of GMOs are diverse and include drugs in food, bananas that produce human vaccines against infectious diseases such as Hepatitis B, metabolically engineered fish that mature more quickly, fruit and nut trees that yield years earlier, and plants that produce new plastics with unique properties. While their practicality or efficacy in commercial production has yet to be fully tested, the next decade may see exponential increases in GM product development as researchers gain increasing access to genomic resources that are applicable to organisms beyond the scope of individual projects.

Safety testing of these products will also at the same time be necessary to ensure that the perceived benefits will indeed outweigh the perceived and hidden costs of development. Controversies surrounding GM foods and crops commonly focus on human and environmental safety, labeling and consumer choice, intellectual property rights, ethics, food security, poverty reduction, and environmental conservation. See also: GM food controversy

Safety Testing

In the USA regulation of a genetically modified food is determined by the objective characteristics of the food and the intended use of the food, irrespective of the way it was developed. FDA policy states that a formal pre-market review by the FDA is to be taken when the objective characteristics of any substance added to the food raises safety issues.

Prior to marketing a new GM food product, manufacturers are required to submit documentation to the FDA to demonstrate its safety and then await approval before selling it to consumers. The context for assessing safety of novel foods is the fact that existing foods often contain toxic components but are still able to be consumed safely. For instance, potatoes and tomatoes can contain toxic levels of solanine and alpha-tomatine alkaloids respectively, and this situation is recognised in the concept of "Substantial Equivalence" that was developed by the OECD in 1993 as a criterion for identifying whether a novel food is at least as safe as the equivalent existing food. The US FDA takes a safety assessment approach that is consistent with this OECD concept in their regulation of novel foods (including those made by recombinant DNA methods). This policy is outlined in an FDA statement.

Critics of GM food believe this regulatory model fails to sufficiently protect consumers and claim that the FDA is subject to pressure and influence by industry. One concern voiced is that a novel crop may have unintended changes created during the insertion of new genetic material. On the other hand, plant scientists, backed by results of modern comprehensive profiling of crop composition, point out that crops modified using GM techniques are less likely to have unintended changes than are conventionally bred crops.

Intellectual Property

Enforcement of Patents on genetically modified plants is often contentious, especially because of the occurrence of Gene flow. In 1998, 95-98% of about 10 km^2 planted with canola by Canadian farmer Percy Schmeiser were found to contain Monsanto's patented Roundup Ready gene although Schmeiser had never purchased seed from Monsanto. The initial source of the plants was undetermined, and could have been through either gene flow or intentional theft. However, the overwhelming predominance of the trait implied that Schmeiser must have intentionally selected for it.

The court determined that Schmeiser had saved seed from areas on and adjacent to his property where Roundup had been sprayed, such as ditches and near power poles. Although unable to prove direct theft, Monsanto sued Schmeiser for piracy since he knowingly grew Roundup Ready plants without paying royalties.

The case made it to the Canadian Supreme Court, which in 2004 ruled 5 to 4 in Monsanto's favour. The dissenting judges focused primarily on the fact that Monsanto's patents covered only the gene itself and glyphosate resistant *cells*, and failed to cover transgenic plants in their entirety. In response to criticism, Monsanto Canada's director of public affairs stated that "It is not, nor has it ever been Monsanto Canada's policy to enforce its patent on Roundup Ready crops when they are present on a farmer's field by accident...Only when there has been a knowing and deliberate violation of its patent rights will Monsanto act." Currently Percy Schmeiser spends a large amount of his time traveling and speaking about how Monsanto ruined his career as a farmer. He also talks about the possible harms of genetically modified and why others in addition to himself should be protesting it.

Benefits and Controversies

Some argue that there is more than enough food in the world and that the hunger crisis is caused by problems in food distribution and politics, not production, so people should not be offered food that may carry some degree of risk. Others oppose genetic engineering on the grounds that genetic modifications might have unforeseen consequences, both in the initially modified organisms and their environments. For example, certain strains of maize have been developed that are toxic to plant eating insects. It has been alleged those strains cross-pollinated with other varieties of wild and domestic maize and passed on these genes with a putative impact on Maize biodiversity.

Subsequent to the publication of these results, several scientists pointed out that the conclusions were based on experiments with design flaws. It is well known that the results from the Polymerase Chain Reaction method of analysing DNA can often be confounded by sample contamination and experimental artifacts. Appropriate controls can be included in experiments to eliminate these as a possible explanation of the results - however these controls were not included in the methods used by Quist and Chapela.

After this criticism *Nature*, the scientific journal where this data was originally published concluded that "the evidence available is not sufficient to justify the publication of the original paper". More recent attempts to replicate the original studies have concluded that genetically modified corn is absent from southern Mexico in 2003 and 2004. Also in dispute is the impact on biodiversity of the introgression of transgenes into wild populations. Unless a transgene offers a massive selective advantage in a wild population, a transgene that enters such a population will be maintained at a low gene frequency. In such situations it can be argued that such an introgression actually *increases* biodiversity rather than lowers it.

Activists opposed to genetic engineering say that with current recombinant technology there is no way to ensure that genetically modified organisms will remain under control, and the use of this technology outside secure laboratory environments carries potentially unacceptable risks to both farmed and wild ecosystems.

Potential impact on biodiversity may occur if herbicide-tolerant crops are sprayed with herbicide to the extent that no wild plants ('weeds') are able to survive. Plants toxic to insects may mean insect-free crops. This could result in declines in other wildlife (e.g. birds) which feed on weed seeds and/or insects for food resources. The recent (2003) farm scale studies in the UK found this to be the case with GM sugar beet and GM rapeseed, but not with GM maize (though in the last instance, the non-GM comparison maize crop had also been treated with environmentally-damaging pesticides subsequently (2004) withdrawn from use in the EU).

Although some scientists have claimed that selective breeding is a form of genetic engineering, (e.g., maize was modified from teosinte, dogs have evolved with human intervention over the course of tens of thousands of years from wolves), others assert that modern transgenesis-based genetic engineering is capable of delivering changes faster than, and sometimes of different types from, traditional breeding methods.

Proponents of current genetic techniques as applied to food plants cite the benefits that the technology can have, for example, in the harsh agricultural conditions of Africa. They say that with modifications, existing crops would be able to thrive under the relatively hostile conditions providing much needed food to their people. Proponents also cite golden rice and golden rice 2, genetically engineered rice varieties (still under development) that contain elevated vitamin A levels. There is hope that this rice may alleviate vitamin A deficiency that contributes to the death of millions and permanent blindness of 500,000 annually.

Proponents say that genetically-engineered crops are not significantly different from those modified by nature or humans in the past, and are as safe or even safer than such methods. There is gene transfer between unicellular eukaryotes and prokaryotes. There have been no known genetic catastrophes as a result of this. They argue that animal husbandry and crop breeding are also forms of genetic engineering that use artificial selection instead of modern genetic modification techniques. It is politics, they argue, not economics or science, that causes their work to be closely investigated, and for different standards to apply to it than those applied to other forms of agricultural technology.

Proponents also note that species or genera barriers have been crossed in nature in the past. An oft-cited example is today's modern red wheat variety, which is the result of two natural crossings made long ago. It is made up of three groups of seven chromosomes. Each of those three groups came from a different wild wheat grass. First, a cross between two of the grasses occurred, creating the durum wheats, which were the commercial grains of the first civilizations up through the Roman Republic. Then a cross occurred between that 14-chromosome durum wheat and another wild grass to create what became modern red wheat at the time of the Roman Empire.

Economic and Political Effects

- Many opponents of current genetic engineering believe the increasing use of GM in major crops has caused a power shift in agriculture towards Biotechnology companies, which are gaining excessive control over the production chain of crops and food, and over the farmers that use their products, as well.
- Many proponents of some current genetic engineering techniques believe it will lower pesticide usage and has brought higher yields and profitability to many farmers, including those in developing nations. A few genetic engineering licenses allow farmers in less economically developed countries to save seeds for next year's planting.
- In August 2002, Zambia cut off the flow of Genetically Modified Food (mostly maize) from UN's World Food Programme. Although there were claims that this left a famine-stricken population without food aid, the U.N. programme succeeded in replacing the rejected grain with other sources, including some foods purchased locally with European cash donations. In rejecting the maize, Zambians cited the "Precautionary Principle" and also the desire to protect future possibilities of grain exports to Europe.
- In December 2005 the Zambian government changed its mind in the face of further famine and allowed the importation of GM maize. However, the Zambian Minister for Agriculture Mundia Sikatana has insisted that the ban on genetically modified maize remains, saying "We do not want GM (genetically modified) foods and our hope is that all of us can continue to produce non-GM foods."
- In April 2004 Hugo Chavez announced a total ban on genetically modified seeds in Venezuela.
- In January 2005, the Hungarian government announced a ban on importing and planting of genetic modified maize seeds, although these were agreed authorized by the EU.
- On August 18, 2006, American exports of rice to Europe were interrupted when much of the U.S. crop was confirmed to be contaminated with unapproved engineered genes, possibly due to

accidental cross-pollination with conventional crops. The U.S. government has since declared the rice safe for human consumption, and exports to some countries have since resumed.

Compared with traditional cross breeding techniques where hundreds and thousands of genes are transferred indiscriminately, using genetic engineering techniques to develop GM crops with the introduction of one or a few genes results in subtle and less disruptive changes that are relatively specific and predictable.

Direct gene transfer is still a relatively new procedure and can be viewed as a logical extension to the methods that have been used for thousands of years to modify our crop plants. Even so, many questions have arisen concerning the safety of these foods and their potential impact on the environment.

Safety of Biotech Foods

The first genetically engineered whole product available to the marketplace was the Flavr Savr. tomato, introduced to the public in May 1994 without much fanfare. The Flavr Savr. tomato was genetically engineered by Calgene, Inc., to stay ripe without rotting for up to ten days, allowing plenty of time for the vine-ripened tomato to be picked ripe, shipped, and sold. Calgene, although not required to, sought approval by the FDA to assure consumers that their product was safe for consumption.

Since the approval of the Flavr Savr. tomato, every developer of biotech foods has, without exception, consulted with the FDA on a voluntary basis to preemptively address any safety or nutrition related concerns before the product reaches the marketplace. In the eight years since the introduction of the Flavr Savr. tomato, there hasn't been a single documented case of an illness caused by biotech foods.

In the U.S., the safety of biotech foods is overseen by three separate agencies:

- The Food and Drug Administration (FDA) assesses the safety of all foods and animal feeds, including those produced through plant biotechnology.
- The Department of Agriculture (USDA), through its Animal and Plant Health Inspection Service, oversees field-testing of biotech seeds and plants to make sure their release causes no harm to the environment, especially native plants.
- The Environmental Protection Agency (EPA) evaluates biotech plants' environmental safety such as their pesticide properties, possible effect on wildlife, and how these plants break down in the environment.

Issues of concern have also been raised with regard to biotech foods and the environment. Cornell University researchers found, while conducting laboratory studies, that monarch larvae could be harmed or killed if they ate large amounts of Bt corn pollen. When Cornell University researchers conducted their 1999 study they did not attempt to duplicate real-world environmental conditions.

They used only a small number of caterpillars and gave them no choice but to eat leaves coated with a thick layer of Bt corn pollen. Subsequent field studies conducted by the USDA's Agriculture Research Service (ARS) demonstrated that Bt corn does not harm the monarch butterfly. Extensive studies from around the world show that biotech crops are helping to preserve and protect the environment.

Researchers from Washington State University, the University of Illinois, Clemson University, and the National Centre for Food and Agricultural Policy (NCFAP) found that through the use of biotech crops, the following is possible:

- Valuable soil erosion can be prevented using environmentally friendly no-till farming practices when planting biotech soybeans and cotton.
- The ability to use more benign herbicides that rapidly dissipate in soil and water improves water quality.
- Air quality is improved when no-till farming practices are used, significantly reducing the release of greenhouse gasses.

There are also concerns that eating GM foods could lead to an increase of unintentional allergic reactions. Extensive testing is part of the approval process to make sure that proteins introduced to crops to improve their characteristics do not cause new allergies. Scientists believe that biotechnology is a promising tool for removing allergens from foods, giving many allergy-prone people a wider choice of safe foods to eat.

Even before they reach a farmer's field, biotech corn, soybeans, and other genetically enhanced foods undergo years of review by researchers, university scientists, farmers, and other government agencies in addition to the FDA. The results of these tests have all produced the same conclusions. Biotech crops are safe to eat. A report issued by the National Academy of Sciences, an independent group of scientists and scholars, confirmed that all approved biotech products are as safe as their conventional counterparts for human and animal consumption.

Benefits of GM Foods

With a continuously expanding world population that already exceeds six billion people, ensuring an adequate food supply is going to become a major challenge. GM foods can potentially help to meet this challenge in a number of ways:

- Pest resistant crops—world crop productivity could increase by as much as 25 percent through the use of biotechnology to grow plants that resist pests and diseases. Growing GM foods such as Bt corn can help eliminate the application of chemical pesticides, reducing the run-off of agriculture wastes that can poison valuable water sources and cause harm to the environment.
- Disease resistant crops—scientists are creating plants that have the ability to fight against many viruses, fungi, and bacteria that are harmful to them.
- Cold tolerance—by introducing an antifreeze gene from cold water fish, plants are being developed that are able to withstand cold temperatures that would normally kill them.
- Drought tolerance/salinity tolerance—plants created to withstand periods of drought or high salt content in their soil and groundwater will help to grow crops in land formerly unsuitable for growing.
- Pharmaceuticals—vaccines introduced into tomatoes and potatoes are much easier to ship, store, and administer than traditional vaccines that have to be injected, a major benefit for developing third-world countries.

International food and agricultural experts and policy makers—including the U.N. Food and Agricultural Organization and the World Health Organization—call plant biotechnology a critical tool to help feed a growing population in the twenty-first century. Using existing farmland, world crop productivity could increase by as much as 25 percent through the use of GM crops engineered to resist pests and diseases, tolerate harsh growing conditions, and delay ripening to reduce spoilage.

A study conducted by the National Centre for Food and Agricultural Policy (NCFAP), released in June 2002, found that six biotech crops planted in the U.S.—soybeans, corn, cotton, papaya, squash, and canola—produced an additional four billion pounds of food and fiber on the same acreage, improved farm income by $1.5 billion, and reduced pesticide use by 46 million pounds.

While some nations vacillate about the merits of genetically modified crops, other nations such as China, Africa, India, and Pacific Island nations embrace GM crops as a means to expand production of food crops in order to feed the world's fastest-growing populations. Rice is a staple crop for nearly a billion people in Asian countries where 50 percent or more of calories are from rice. Aside from providing more calories from higher yields, genetically modified rice can provide increased nutrients such as vitamin A, lysine, iron, and zinc, thus reducing illnesses and malnutrition associated with deficiencies of these nutrients.

Labeling Concerns

The labeling of GM foods and food products is a controversial issue. Food industries support the belief that labeling should be voluntary and influenced by consumer demands and say that requiring a label for biotech ingredients would confuse consumers more than it would inform them. Consumer interest groups are demanding mandatory labeling, contending that people have the right to know what they are eating.

The Food and Drug Administration (FDA), which oversees food safety issues, performs exhaustive safety tests on every biotech food product entering the marketplace and has concluded that special labeling is required only when the new food product is significantly different from its conventional counterpart. The American Medical Association (AMA) has also stated that, "there is no scientific justification for special labeling of genetically modified foods, as a class".

Special labeling requirements for GM foods would dramatically increase the costs of all foods—costs that would be passed on to the consumer. The FDA is currently developing voluntary labeling guidelines, which would enable manufacturers to let consumers know if a food was developed using biotechnology to have a beneficial trait, or if biotech ingredients were used in making a product. Any type of food labeling should be designed to clearly convey accurate information about the product in language that everyone can understand. The biggest challenge in any GM food labeling policy would be how to educate and inform the public without causing undue alarm or fear of GM foods.

Each year the world's population increases by approximately 73 million people and is projected to reach approximately eight billion by the year 2030—with most of the increase expected to occur in developing countries. Despite efforts worldwide to combat hunger and malnutrition, there are over 800 million people worldwide who do not receive adequate food and nutrition. In order for the world to address this critical situation, the world will have to double its food production and distribution over the next 25 years.

Plant biotechnology can help to overcome the world's concern for feeding its ever-growing population safely and effectively by continuing to develop a food supply that is more abundant on less acreage, has more nutritional value, and is more environmentally conservative.

Governments around the world need to develop policies to ensure greater investment in research and at the same time develop necessary regulations and guidelines that will enable continued advances in plant biotechnology while maintaining confidence and trust that our food supply is safe for consumption and the ecosystem of our world.

Standards for Technological Literacy

Introducing students to plant biotechnology will expose students to the relatively new science of genetically engineered foods and help them to understand the concerns, controversy, and potential that this technology espouses. As more and more foods are introduced to our daily diet, it is important that students understand the methods used to develop these foods, regulations, and guidelines used to determine their safety for human consumption, and tests that are performed to ensure that our environment is not harmed.

Standards for Technological Literacy offers benchmark guides and suggestions that will help technology education teachers in their preparation to expose their students to the wonders and potential of plant biotechnology.

Genetic Engineering Activity: Regulation and Policies for Genetically Modified Crops

Objectives:

- Students will become informed about the processes used to genetically modify crops.
- Students will assess the impact of genetically modified crops on society and the environment and make intelligent decisions about their value and how they should be regulated.

As we have seen, genetic engineering has the potential to alter plant and animal life forms in ways not found in nature. For centuries, farmers have used selective breeding to improve both crops and animals that had the characteristics they wanted to bring out and strengthen. This was the only way that farmers had to develop plants and animals that were resistant to disease and tolerant of climate extremes.

Today, scientists can find individual genes that control particular characteristics, separate them out, change them, and transfer them directly into the cells of an animal, plant, bacterium, or virus. Because the DNA code is known and is common to all life, it is also possible to produce synthetic genes. This technology is called genetic modification or genetic engineering.

Genetic modification means that, for the first time, humans can make living things to our own design, without relying on nature. The implications for this new technology are vast. Although any new technology may have its risks and benefits, genetic engineering has special features that need to be addressed with wisdom and insight. There is much controversy about genetically engineered or modified plants such as Bt corn (bacillus thuringiensis), Bt cotton, and soybeans. One side is that genetically modified crops provide benefits such as increased crop yields, reduced pesticide use, and consistent quality under varying growing conditions, and improved economic returns. Additionally, GM crops could help meet the nutritional needs of millions of people around the globe who are literally starving.

Conversely, there are risks in developing and releasing GM crops for food consumption before they are thoroughly tested. The media recently reported that a public interest group found StarLink's genetically modified protein in taco shells, sparking contentious international debate over crop biotechnology.

The claim was made that the Starlink[R] contained an allergen that may affect some people. Subsequently, Starlink[R] has been banned for human consumption. Could a GM plant cross-pollinate with another plant to produce a high-risk offspring? These issues and others present real challenges to companies developing GM plants as well as to governmental and regulatory agencies and consumers here and abroad.

As a student of technology, you are aware of the assessment process used to evaluate existing and emerging technologies. Given the fact that GM crops are an invention rather than a product of nature and natural selection, it would be appropriate to assess the benefits of GM crops and propose a set of policies that could be used to regulate them. It is important to note that care must be used in establishing policies that will encourage the development of new GM crops, yet protect the public and environment. Policies that are too restrictive will inhibit innovation and development and become a barrier to the introduction of GM crops. Policies that are ineffective may result in GM crops that could create irreversible impacts on the ecological environment.

The following are issues that can provide a starting point for developing a set of policies regarding genetically engineered crops.

- Benefit to humans—nutritional and economic
- Human consumption risks—allergic and intestinal reactions
- Intellectual property-patents, ownership, licensing, and royalties
- Environmental risks—effects on other plants and insects
- Ecological risks
- Economic impacts
- Safety and testing
- Control and monitoring the distribution of GM seed and produce
- Liability, compensation, and risk management
- Should genetically modified foods be labeled?

Plant Breeding

Plant breeding is the purposeful manipulation of plant species in order to create desired genotypes and phenotypes for specific purposes. This manipulation involves either controlled pollination, genetic engineering, or both, followed by artificial selection of progeny. *Plant breeding* often, but not always, leads to plant domestication.

Plant breeding has been practiced for thousands of years, since near the beginning of human civilization. It is now practiced worldwide by government institutions and commercial enterprises. International development agencies believe that breeding new crops is important for ensuring food security and developing practices of sustainable agriculture through the development of crops suitable for their environment.

Domestication

Domestication of plants is an artificial selection process conducted by humans to produce plants that have fewer undesirable traits of wild plants, and which renders them dependent on artificial (usually enhanced) environments for their continued existence. The practice is estimated to date back 9,000-11,000 years. Many crops in present day cultivation are the result of domestication in ancient times, about 5,000 years ago in the Old World and 3,000 years ago in the New World. In the Neolithic period, domestication took a minimum of 1,000 years and a maximum of 7,000 years. Today, all of our principal food crops come from domesticated varieties.

A cultivated crop species that has evolved from wild populations due to selective pressures from traditional farmers is called a landrace. Landraces, which can be the result of natural forces or domestication, are plants (or animals) that are ideally suited to a particular region or environment. Examples are the landraces of rice, *Oryza sativa* subspecies *indica,* which was developed in South Asia, and *Oryza sativa* subspecies *japonica,* which was developed in China.

Classical Plant Breeding

Classical plant breeding uses deliberate interbreeding (crossing) of closely or distantly related individuals to produce new crop varieties or lines with desirable properties. Plants are crossbred to introduce traits/genes from one variety or line into a new genetic background. For example, a mildew-resistant pea may be crossed with a high-yielding but susceptible pea, the goal of the cross being to introduce mildew resistance without losing the high-yield characteristics. Progeny from the cross would then be crossed with the high-yielding parent to ensure that the progeny were most like the high-yielding parent, (backcrossing). The progeny from that cross would then be tested for yield and mildew resistance and high-yielding resistant plants would be further developed. Plants may also be crossed with themselves to produce inbred varieties for breeding.

Classical breeding relies largely on homologous recombination of two genomes to generate genetic diversity. The classical plant breeder may also makes use of a number of *in vitro* techniques such as protoplast fusion, embryo rescue or mutagenesis (see below) to generate diversity and produce Transgenic plants that would not exist in nature.

The Yecoro wheat (right) cultivar is sensitive to salinity, plants resulting from a hybrid cross with cultivar W4910 (left) show greater tolerance to high salinity

Traits that breeders have tried to incorporate into crop plants in the last 100 years include:

1. Increased quality and yield of the crop
2. Increased tolerance of environmental pressures (salinity, extreme temperature, drought)
3. Resistance to viruses, fungi and bacteria
4. Increased tolerance to insect pests
5. Increased tolerance of herbicides

Intraspecific hybridization within a plant species was demonstrated by Charles Darwin and Gregor Mendel, and was further developed by geneticists and plant breeders. In the early 20th century, plant breeders realized that Mendel's findings on the non-random nature of inheritance could be applied to seedling populations produced through deliberate pollinations to predict the frequencies of different types.

In 1908, George Harrison Shull described heterosis, also known as hybrid vigour. Heterosis describes the tendency of the progeny of a specific cross to outperform both parents. The detection of the usefulness of heterosis for plant breeding has lead to the development of inbred lines that reveal a heterotic yield advantage when they are crossed. Maize was the first species where heterosis was widely used to produce hybrids.

By the 1920s, statistical methods were developed to analyze gene action and distinguish heritable variation from variation caused by environment. In 1933, another important breeding technique, cytoplasmic male sterility (CMS), developed in maize, was described by Marcus Morton Rhoades. CMS is a maternally inherited trait that makes the plant produce sterile pollen, enabling the production of hybrids and removing the need for detasseling maize plants.

These early breeding techniques resulted in large yield increase in the United States in the early 20th century. Similar yield increases were not produced elsewhere until after World War II, the Green Revolution increased crop production in the developing world in the 1960s. Following World War II a number of techniques were developed that allowed plant breeders to hybridize distantly related species, and artificially induce genetic diversity.

When distantly related species are crossed, plant breeders make use of a number of plant tissue culture techniques to produce progeny from otherwise fruitless mating. Interspecific and intergeneric hybrids are produced from a cross of related species or genera that do not normally sexually reproduce with each other. These crosses are referred to as *Wide crosses*. For example, the cereal triticale is a wheat and rye hybrid.

The cells in the plants derived from the first generation created from the cross contained an uneven number of chromosomes and as result was sterile. The cell division inhibitor colchicine was used to double the number of chromosomes in the cell and thus allow the production of a fertile line.

Failure to produce a hybrid may be due to pre- or post-fertilization incompatibility. If fertilization is possible between two species or genera, the hybrid embryo may abort before maturation. If this does occur the embryo resulting from an interspecific or intergeneric cross can sometimes be rescued and cultured to produce a whole plant. Such a method is referred to as *Embryo Rescue*. This technique has been used to produce new rice for Africa, an interspecific cross of Asian rice *(Oryza sativa)* and African rice *(Oryza glaberrima)*.

Hybrids may also be produced by a technique called protoplast fusion. In this case protoplasts are fused, usually in an electric field. Viable recombinants can be regenerated in culture.

Chemical mutagens like EMS and DMSO, radiation and transposons are used to generate mutants with desirable traits to be bred with other cultivars. Classical plant breeders also generate genetic diversity within a species by exploiting a process called somaclonal variation, which occurs in plants produced from tissue culture, particularly plants derived from callus. Induced polyploidy, and the addition or removal of chromosomes using a technique called chromosome engineering may also be used.

When a desirable trait has been bred into a species, a number of crosses to the favoured parent are made to make the new plant as similar as the parent as possible. Returning to the example of the mildew resistant pea being crossed with a high-yielding but susceptible pea, to make the mildew resistant progeny of the cross most like the high-yielding parent, the progeny will be crossed back to that parent for several generations. This process removes most of the genetic contribution of the mildew resistant parent. Classical breeding is therefore a cyclical process. It should be noted that with classical breeding techniques, the breeder does not know exactly what genes have been introduced to the new cultivars. Some scientists therefore argue that plants produced by classical breeding methods should undergo the same safety testing regime as genetically modified plants. There have been instances where plants bred using classical techniques have been unsuitable for human consumption, for example the poison solanine was unintentionally increased to unacceptable levels in certain varieties of potato though plant breeding. New potato varieties are often screened for solanine levels before reaching the marketplace.

Modern plant breeding

Modern plant breeding uses techniques of molecular biology to select, or in the case of genetic modification, to insert, desirable traits into plants.

Marker-Assisted Selection

Sometimes many different genes can influence a desirable trait in plant breeding. The use of tools such as molecular markers or DNA fingerprinting can map thousands of genes. This allows plant breeders to screen large populations of plants for those that possess the trait of interest. The screening is based on the presence or absence of a certain gene as determined by laboratory procedures, rather than on the visual identification of the expressed trait in the plant.

Genetic Modification

Genetic modification of plants is achieved by adding a specific gene or genes to a plant, or by knocking out a gene with RNAi, to produce a desirable phenotype. The plants resulting from adding a gene are often referred to as transgenic plants. If for genetic modification genes of the species or of a crossable plant are used under control of their native promoter, then they are called Cisgenic plants. Genetic modification can produce a plant with the desired trait or traits faster than classical breeding because the majority of the plant's genome is not altered.

To genetically modify a plant, a genetic construct must be designed so that the gene to be added or knocked-out will be expressed by the plant. To do this, a promoter to drive transcription and a termination sequence to stop transcription of the new gene, and the gene of genes of interest must be introduced to the plant. A marker for the selection of transformed plants is also included. In the labouratory, antibiotic resistance is a commonly used marker: plants that have been successfully transformed will grow on media containing antibiotics; plants that have not been transformed will die. In some instances markers for selection are removed by backcrossing with the parent plant prior to commercial release. The construct can be inserted in the plant genome by genetic recombination using the bacteria *Agrobacterium tumefaciens* or *A. rhizogenes*, or by direct methods like the gene gun or microinjection. Using plant viruses to insert genetic constructs into plants is also a possibility, but the technique is limited by the host range of the virus. For example, Cauliflower mosaic virus (CaMV) only infects cauliflower and related species. Another limitation of viral vectors is that the virus is not usually passed on the progeny, so every plant has to be inoculated.

The majority of commercially released transgenic plants, are currently limited to plants that have introduced resistance to insect pests and herbicides. Insect resistance is achieved through incorporation of a gene from *Bacillus thuringiensis* (Bt) that encodes a protein that is toxic to some insects. For example, the cotton bollworm, a common cotton pest, feeds on Bt cotton it will ingest the toxin and die. Herbicides usually work by binding to certain plant enzymes and inhibiting their action.

The enzymes that the herbicide inhibits are known as the herbicides *target site*. Herbicide resistance can be engineered into crops by expressing a version of *target site* protein that is not inhibited by the herbicide. This is the method used to produce glyphosate resistant crop plants. Genetic modification of plants that can produce pharmaceuticals (and industrial chemicals), sometimes called *pharmacrops*, is a rather radical new area of plant breeding.

Issues and Concerns

Modern plant breeding, whether classical or through genetic engineering, comes with issues of concern, particularly with regard to food crops. The question of whether breeding can have a negative effect on nutritional value is central in this respect. Although relatively little direct research in this area has been done, there are scientific indications that, by favouring certain aspects of a plant's development, other aspects may be retarded.

A study published in the Journal of the American College of Nutrition in 2004, entitled Changes in USDA Food Composition Data for 43 Garden Crops, 1950 to 1999, compared nutritional analysis of vegetables done in 1950 and in 1999, and found substantial decreases in six of 13 nutrients measured, including 6% of protein and 38% of riboflavin. Reductions in calcium, phosphorus, iron and ascorbic acid were also found. The study, conducted at the Biochemical Institute, University of Texas at Austin, concluded in summary: "We suggest that any real declines are generally most easily explained by changes in cultivated varieties between 1950 and 1999, in which there may be trade-offs between yield and nutrient content." The debate surrounding genetic modification of plants is huge, encompassing the ecological impact of genetically modified plants, the safety of genetically modified food and concepts used for safety evaluation like Substantial equivalence.

Plant breeders' rights is also a major and controversial issue. Today, production of new varieties is dominated by commercial plant breeders, who seek to protect their work and collect royalties through national and international agreements based in intellectual property rights.

The range of related issues is complex. In the simplest terms, critics of the increasingly restrictive regulations argue that, through a combination of technical and economic pressures, commercial breeders are reducing biodiversity and significantly constraining individuals (such as farmers) from developing and trading seed on a regional level. Efforts to strengthen breeders' rights, for example, by lengthening periods of variety protection, are ongoing.

Transgenic Plant

Transgenic plants are plants that possess a gene or genes that have been transferred from a different species. Such modification may be performed through ordinary hybridization through cross-pollination of plants, but the term today refers to plants produced in a labouratory using recombinant DNA technology in order to create plants with specific characteristics by artificial insertion of genes from other species, and sometimes entirely different kingdoms.

Prior to the current era of molecular genetics starting around 1975, transgenic plants, including cereal crops, were (since the mid 1930s) part of conventional plant breeding.

Transgenic varieties are frequently created by classical breeders who deliberately force hybridization between distinct plant species when carrying out interspecific or intergeneric *wide crosses* with the intention of developing disease resistant crop varieties. Classical plant breeder may use use of a number of *in vitro* techniques such as protoplast fusion, embryo rescue or mutagenisis to generate diversity and produce plants that would not exist in nature.

Such traditional techniques (used since about 1930 on) have never been controversial, or been given wide publicity except among professional biologists, and have allowed crop breeders to develop varieties of basic food crop, wheat in particular, which resist devastating plant diseases such as rusts. *Hope* is one such transgenic wheat variety bred by E. S. McFadden with a transgene from a wild grass. *Hope* saved American wheat growers from devastating stem rust outbreaks in the 1930s.

Methods used in traditional breeding that generate transgenic plants by non-recombinant methods are widely familiar to professional plant scientists, and serve important roles in securing a sustainable future for agriculture by protecting crops from pests and helping land and water to be used more efficiently.

Natural Movements of Genes

Natural movement of genes between species, often called horizontal gene transfer or lateral gene transfer, can occur because of gene transfer mediated by natural agent This natural gene movement between species has been widely detected during genetic investigation of various natural Mobile genetic elements, such as transposons, and retrotransposons that naturally transfer to new locations in a genome, and often move to new species host over an evolutionary time scale. There are many types of natural mobile DNAs, and they have been detected abundantly in food crops such as rice.

These various mobile genes play a major role in dynamic changes to chromosomes during evolution, and have often been given whimsical names, such as Mariner, Hobo, Trans-Siberian Express (Transib), Osmar, Helitron, Sleeping Princess, MITE and MULE, to emphasize their mobile and transient behaviour.

Such genetically mobile DNA contitutute a major fraction of the DNA of many plants, and the natural dynamic changes to crop plant chromosomes caused by this natural transgenic DNA mimics many of the features of plant genetic engineering currently pursued in the labouratory, such as using transposons as a genetic tool, and molecular cloning. There is new scientific literature about natural transgenic events in plants, through movement of natural mobile DNAs called MULEs between rice and Setaria millet.

It is becoming clear that natural rearrangements of DNA and generation of transgenes play a pervasive role in natural evolution. Importantly many, if not most, flowering plants evolved by transgenesis - that is, the creation of natural interspecies hybrids in which chromosome sets from different plant species were added together. There is also the long and rich history of transgenic varieties in traditional breeding.

Creation of transgenic plants

Production of transgenic plants in wide-crosses by plant breeders has been a vital aspect of conventional plant breeding for a century or so. Without it, security of our food supply against losses caused by crop pests such as rusts and mildews would be severely compromised. The first historically recorded interpecies transgenic cereal hybrid was actually between wheat and rye. Last century, the introduction of alien germplasm into common foods was repeatedly achieved by traditional crop breeders by artificially overcoming fertility barriers. A novel genetic rearrangement of plant chromosomes, such as insertion of large blocks of rye (Secale) genes into wheat chromosomes ('translocations'), has also been exploited widely for many decades.By the late 1930s with the introduction of colchicine, perennial grasses were being hybridized with wheat with the aim of transferring disease resistance and perenniality into annual crops, and large-scale practical use of hybrids was well established, leading on to development of Triticosecale and other new transgenic cereal crops. In 1985 Plant Genetic Systems (Ghent, Belgium), founded by Marc Van Montagu and Jeff Schell, was the first company to develop genetically engineered (tobacco) plants with insect tolerance by expressing genes encoding for insecticidal proteins from Bacillus thuringiensis (Bt). Important transgenic pathogen and parasite resistance traits in current bread wheat varieties (gene, eg "Lr9" followed by the source species) are:

Disease resistance

- Leaf rust
  - Lr9 (from *Aegilops umbellulata*)
  - Lr18 *Triticum timopheevi*
  - Lr19 *Thinopyrum*
  - Lr23 *T. turgidum*
  - Lr24 *Ag. elongatum*
  - Lr25 *Secale cereale*
  - Lr29 *Ag. elongatum*
  - Lr32 *T. tauschii*
- Stem rust
  - Sr2 *T. turgidum* ("Hope") McFadden, E. S., J. Am. Soc. Agron. 22, 1020-1031.
  - Sr22 *Triticum monococcum*
  - Sr36 *Triticum timopheevii*
- Stripe rust
  - Yr15 *Triticum dicoccoides*
- Powdery mildew
  - Pm12 *Aegilops speltoides*
  - Pm21 *Haynaldia villosa*
  - Pm25 *T. monococcum*
- Wheat streak mosaic virus
  - Wsm1 *Ag. elongatum*

Pest resistance

- Hessian fly
  - H21 *S. cereale* H23,
  - H24 *T. tauschii*
  - H27 *Aegilops ventricosa*
- Cereal cyst nematode
  - Cre3 (Ccn-D1) *T. tauschii*
- Lepidoptera
  - Bt *Bacillus thuringiensis*

The intentional creation of transgenic plants by laboratory based recombinant DNA methods is more recent (from the mid-80s on) and has been a controversial development opposed vigorously by many NGOs, and several governments, particularly within the European Community. These transgenic recombinant plants (= biotech crops, modern transgenics) are transforming agriculture in those regions that have allowed farmers to adopt them, and the area sown to these crops has continued to grow globally in each of the ten years since their first introduction in 1996.

Transgenic recombinant plants are now generally produced in a laboratory by adding one or more genes to a plant's genome, and the techniques frequently called transformation. Transformation is usually achieved using gold particle bombardment or a soil bacterium (*Agrobacterium tumefaciens*) carrying an engineered plasmid vector, or carrier of selected extra genes. Transgenic recombinant plants are identified as a class of genetically modified organism (GMO); usually only transgenic plants created by direct DNA manipulation are given much attention in public discussions.

Transgenic plants have been deliberately developed for a variety of reasons: longer shelf life, disease resistance, herbicide resistance, pest resistance, non-biological stress resistances, such as to drought or nitrogen starvation, and nutritional improvement. The first modern transgenic crop approved for sale in the US, in 1994, was the FlavrSavr tomato, which was intended to have a longer shelf life. The first conventional transgenic cereal created by scientific breeders was actually a hybrid between wheat and rye in 1876. The first transgenic cereal may have been wheat, which itself is a natural transgenic plant derived from at least three different parenteral species. Commercial factors, especially high regulatory and research costs, have so far restricted modern transgenic crop varieties to major traded commodity crops, but recently R&D projects to enhance crops that are locally important in developing counties are being pursued, such as insect protected cow-pea for Africa and insect protected Brinjal eggplant for India.

Regulation of Transgenic Plants

In the United States the Coordinated Framework for Regulation of Biotechnology governs the regulation of transgenic organisms, including plants. The three agencies involved are:

- USDA Animal and Plant Health Inspection Service

The Biotechnology Regulatory Services (BRS) programme of the U.S. Department of Agriculture's (USDA) Animal and Plant Health Inspection Service (APHIS) is responsible for regulating the introduction (importation, interstate movement, and field release) of genetically engineered (GE) organisms that may pose a plant pest risk. BRS exercises this authority through APHIS regulations in Title 7, Code of Federal Regulations, Part 340 under the Plant Protection Act of 2000. APHIS protects agriculture and the environment by ensuring that biotechnology is developed and used in a safe manner.

Through a strong regulatory framework, BRS ensures the safe and confined introduction of new GE plants with significant safeguards to prevent the accidental release of any GE material. APHIS has regulated the biotechnology industry since 1987 and has authorized more than 10,000 field tests of GE organisms. In order to emphasize the importance of the programme, APHIS established BRS in August 2002 by combining units within the agency that dealt with the regulation of biotechnology.

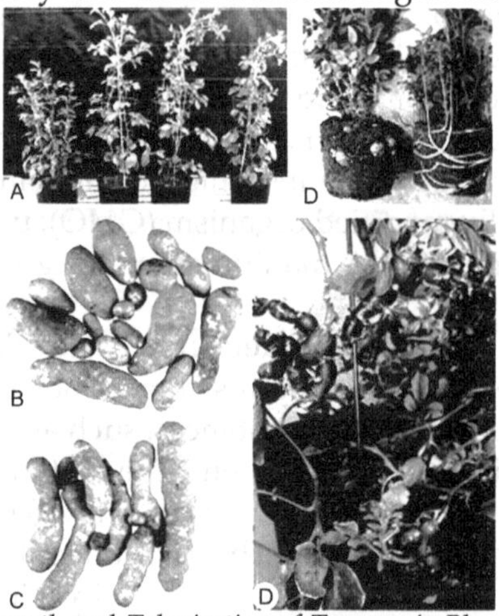

Fig. *Growth and Tuberization of Transgenic Plants with Altered Expression PCM-1 as Compared to Control. A: Control Plant on the left, and three independent Transgenic Plants to the right of control. Comparison of tuber shape and size B: (control), C: (transgenic). D: Tuberization pattern of one of the independent Transgenic Plants expressing the highest amount of PCM-1 (left, control; right, Transgenic plant). E: When the Transgenic Plant from figure D, was allowed to grow, aerial tubers were produced.*

Biotechnology, Federal Regulation, and the U.S. Department of Agriculture, February 2006, USDA-APHIS Fact Sheet

- EPA - evaluates potential environmental impacts, especially for genes which encode for pesticide production
- DHHS, Food and Drug Administration (FDA) - evaluates human health risk if the plant is intended for human consumption

Ecological risks

The potential impact on nearby ecosystems is one of the greatest concerns associated with transgenic plants. Transgenes have the potential for significant ecological impact if the plants can increase in frequency and persist in natural populations. These concerns are similar to those surrounding conventionally bred plant breeds. Several risk factors should be considered:

- Is the transgenic plant capable of growing outside a cultivated area?
- Can the transgenic plant pass its genes to a local wild species, and are the offspring also fertile?
- Does the introduction of the transgene confer a selective advantage to the plant or to hybrids in the wild?

Many domesticated plants can mate and hybridise with wild relatives when they are grown in proximity, and whatever genes the cultivated plant had can then be passed to the hybrid. This applies equally to transgenic plants and conventionally bred plants, as in either case there are advantageous genes that may have negative consequences to an ecosystem upon release. This is normally not a significant concern, despite fears over 'mutant superweeds' overgrowing local wildlife: although hybrid plants are far from uncommon, in most cases these hybrids are not fertile due to polyploidy, and will not multiply or persist long after the original domestic plant is removed from the environment. However, this does not negate the possibility of a negative impact.

In some cases, the pollen from a domestic plant may travel many miles on the wind before fertilising another plant. This can make it difficult to assess the potential harm of crossbreeding; many of the relevant hybrids are far away from the test site. Among the solutions under study for this concern are systems designed to prevent transfer of transgenes, such as Terminator. Technology, and the genetic transformation of the chloroplast only, so that only the seed of the transgenic plant would bear the transgene. With regard to the latter, there is some controversy that the technologies may be inequitable and might force dependence upon producers for valid seed in the case of poor farmers, whereas the latter has no such concern but has technical constraints that still need to be overcome. There are at least three possible avenues of hybridization leading to escape of a transgene:

- Hybridization with non-transgenic crop plants of the same species and variety.
- Hybridization with wild plants of the same species.
- Hybridization with wild plants of closely related species, usually of the same genus.

However, there are a number of factors which must be present for hybrids to be created.

- The transgenic plants must be close enough to the wild species for the pollen to reach the wild plants.
- The wild and transgenic plants must flower at the same time.
- The wild and transgenic plants must be genetically compatible.

In order to persist, these hybrid offspring:

- Must be viable, and fertile.
- Must carry the transgene.

Studies suggest that a possible escape route for transgenic plants will be through hybridization with wild plants of related species.

1. It is known that some crop plants have been found to hybridize with wild counterparts.
2. It is understood, as a basic part of population genetics, that the spread of a transgene in a wild population will be directly related to the fitness effects of the gene in addition to the rate of influx of the gene to the population. Advantageous genes will spread rapidly, neutral genes will spread with genetic drift, and disadvantageous genes will only spread if there is a constant influx.
3. The ecological effects of transgenes are not known, but it is generally accepted that only genes which improve fitness in relation to abiotic factors would give hybrid plants sufficient advantages to become weedy or invasive. Abiotic factors are parts of the ecosystem which are not alive, such as climate, salt and mineral content, and temperature. Genes improving fitness in relation to biotic factors could disturb the (sometimes fragile) balance of an ecosystem. For instance, a wild plant receiving a pest resistance gene from a transgenic plant might become resistant to one of its natural pests, say, a beetle. This could allow the plant to increase in frequency, while at the same time animals higher up in the food chain, which are at least partly dependent on that beetle as food

source, might decrease in abundance. However, the exact consequences of a transgene with a selective advantage in the natural environment are almost impossible to predict reliably.

It is also important to refer to the demanding actions that government of developing countries had been building up among the last decades.

Genetically modified food controversies

The GM food controversy is a dispute over the advantages and disadvantages of genetically modified food crops. Although no major health hazards have come to light since GM food was introduced 13 years ago, and close to 150 studies are published to attest their safety. Consumer rights groups such as the Organic Consumers Association and Greenpeace emphasize the long term health risks which GM could pose, or that the risks of GM have not yet been adequately investigated.

Unnecessary delays to GM crop use by farmers pose another kind of risk. Agricultural scientist and economists express concern about the harm delaying welfare and environmental improvements, for instance by pro-vitamin A enriched Golden rice which has the potential to prevent much childhood death from infectious disease, and insect protected Bt rice which can reduce exposure of farmers to synthetic insecticides.

Safety disputes

Toxic GM-potatoes

In August 1998 widespread concern, especially in Europe, was sparked by remarks by nutrition researcher, Dr Arpad Pusztai, regarding some of his research into the safety of GM foods.

Pusztai claimed his experiments showed that rats fed on potatoes genetically engineered to express a lectin from snowdrop had suffered serious damage to their immune systems and shown stunted growth. The lectin expressed by the genetically modified potatoes is toxic to insects and nematodes and is allegedly toxic to mammals. He was criticized by leading British politicians, the majority of scientific peers with expertise in the area and by the GM companies because the announcement of his results in a television interview, preceded the scientific publication of his results. When his studies were finally published in The Lancet, no evidence of stunted growth or damage to immune system was substantiated. The Lancet paper's actual summary was:

Diets containing genetically modified (GM) potatoes expressing the lectin *Galanthus nivalis* agglutinin (GNA) had variable effects on different parts of the rat gastrointestinal tract. Some effects, such as the proliferation of the gastric mucosa, were mainly due to the expression of the GNA transgene. However, other parts of the construct or the genetic transformation (or both) could also have contributed to the overall biological effects of the GNA-GM potatoes, particularly on the small intestine and caecum.

The paper was accompanied by an editorial explanation for allowing the paper's publication (Genetically modified foods: "absurd" concern or welcome dialogue?), and an independent critique which had a contrary evaluation of the data: Adequacy of methods for testing the safety of genetically modified foods. This was followed by a lively follow up debate in several later issues of the journal. Nonetheless, controversy about Pusztai's assertions still lingers, caused by strongly held but opposing views on his conclusions and data.

On the one hand, there are claims of misrepresentation of Pusztai's results by Rowett Research Institute, but on the other hand, there are concerns by scientists about overstatement of the quality of his findings by non-governmental organizations, and emphasis on matters well removed from the actual labouratory observations which are rarely discussed in public debate, against the context well over one hundred other studies published by 2006 that support the safety of GM foods and feeds, and commentaries such as that of Nina Fedoroff.

Research protocols were sent by Pusztai to 24 independent scientists in different countries (including experts in physiology, medicine, toxic pathology, nutrition, microbiology and biochemistry). These disagree with the conclusions of the review committee and argued that his research was of good quality and justified his conclusions. Among 'casualties' in these events was Dr Andrew Chesson, vice chairman of European Commission scientific committee on animal nutrition and former top scientist at the Rowett Institute who was fired for publicly defending Pusztai's research.

Various reports concerning the politicisation of the peer review process and alleged deliberate misrepresentation of Pusztai's results were voiced by newspapers and some scientists.

Dangerous Corn

Another controversy recently arose around biotech company Monsanto's data on a 90-Day Rat Feeding Study on a strain of GM corn. In May 2005, critics of GM foods pointed to differences in kidney size and blood composition found in this study, suggesting that the observed differences raises questions about the regulatory concept of substantial equivalence. Some argued that this study suggested human health might be affected by eating GM food. However, the EU regulatory authorities that examined the Monsanto data concluded that the observed small numerical decrease in rat kidney weights was not biologically meaningful, and the weights were well within the normal range of kidney weights for control animals. There were no corresponding microscopic findings in the relevant organ systems, and all blood chemistry and organ weight values fell within the "normal range of historical control values" for rats.

Thus, certain government authorities concluded that there were no effects on the functioning of kidneys in rats fed a diet of GM. Nevertheless, the debate about the Monsanto corn is not over. Genetics Professor Gilles-Eric Seralini, indicated in a scientific study that rats nourished with the ÌÏÍ863 Monsanto corn, showed toxicity signs in their livers and kidneys. The European Committee has approved the ÌÏÍ863 corn for animal and human consumption.

Allergenicity

A gene for an allergenic trait has been transferred unintentionally from the Brazil nut into genetically engineered soybeans while intending to improve soybean nutritional quality for animal feed use. Brazil nuts were already known to produce food allergies in certain people prior to this study. In 1993 Pioneer Hi-Bred International developed a soybean variety with an added gene from the Brazil nut.

This trait increased the levels in the GM soybean of the natural essential amino acid methionine, a protein building block commonly added to poultry feed to improve effective protein quality. Investigation of the GM soybeans revealed that they produced immunological reactions with people suffering from Brazil nut allergy, and the explanation for this is that the methionine rich protein chosen by Pioneer Hi-Bred is the major source of Brazil nut allergy. Pioneer Hi-Bred discontinued further development of the GM soybean and disposed of all material related to the modified soybeans.

While this study indicates the possible risks of GM foods, and indeed any new food source, some point out it establishes the commitment the developmental community has toward consumer safety as well as the competence of current safeguards. Food allergy problems occur with many conventional foods, and Kiwi fruit, for instance, as a relatively new food in many communities, has become widely eaten despite provoking allergies in certain individuals.

Another allergy issue was published in November 2005, when a pest resistant field pea developed by the Australian CSIRO for use as a pasture crop was shown to cause an allergic reaction in mice. Respected plant scientist Maarten J Chrispeels has made interesting comments about this example that illustrate how foods offer many different types of risks: The recent Prescott et al paper in JFAC contains a very interesting study on the immunogenicity of amylase [starch digestion enzyme] inhibitor in its native form (isolated from beans) and expressed as a transgene in peas. First of all, amylase inhibitor is a food protein, but also a "toxic" protein because it inhibits our digestive amylases. This is one of the reasons you have to cook your beans! (The other toxic bean protein is phytohemagglutinin and it is much more toxic).

This particular amylase inhibitor is found in the common bean (other species have other amylase inhibitors). Even though it is a food protein, it is unlikely ever to be used for genetic engineering of human foods because it inhibits our amylases. What the results show is that the protein, when synthesized in pea cotyledons has a different immunogenicity than when it is isolated from bean cotyledons (the native form). This is somewhat surprising but may be related to the presence of slightly different carbohydrate chains.

The immunologist who tested the pea noted that the episode illustrated the need for each new GM food to be very carefully evaluated for potential health effects.

Environmental and Ecological Impacts

As discussed above there is some evidence for positive impacts of the planting of GM crops on reduced greenhouse gas emissions and pesticide loads in the environment. However, there has been controversy over the results of a farm-scale trial in the United Kingdom comparing the impact of GM crops and conventional crops on farmland biodiversity. Some claimed that the results showed that GM crops had a significant negative impact on wildlife.

Others pointed out that the studies showed that using herbicide resistant GM crops allowed better weed control and that under such conditions there were fewer weeds and fewer weed seeds. This result was then extrapolated to suggest that GM crops would have significant impact on the wildlife that might rely on farm weeds. In July 2005 the same British scientists showed that transfer of a herbicide-resistance gene from GM oilseed rape to a wild cousin, charlock, and wild turnips was possible.

Many agricultural scientists and food policy specialists view GM crops as an important element in sustainable food security and environmental management. This point of view is summarised in the ABIC Manifesto: On our planet, 18% of the land mass is used for agricultural production. This fraction cannot be increased substantially. It is absolutely essential that the yield per unit of land increases beyond current levels given that: The human population is still growing, and will reach about nine billion by 2040;70,000 km^2 of agricultural land (equivalent to 60% of the German agricultural area) are lost annually to growth of cities and other non-agricultural uses; Consumer diets in developing countries are increasingly changing from plant-based proteins to animal protein, a trend that requires a greater amount of crop-based feed.

Public Perception

Research by the Pew Initiative on Food and Biotechnology has shown that in 2005 Americans' knowledge of genetically modified foods and animals continues to remain low, and their opinions reflect that they are particularly uncomfortable with animal cloning. The Pew survey also showed that despite continuing concerns about GM foods, American consumers do not support banning new uses of the technology, but rather seek an active role from regulators to ensure that new products are safe.

Interestingly, about 550 Amish farmers in Pennsylvania have adopted GM crops, because they allow for less intensive farming (less pesticides, etc.), are more productive (under these specific conditions), and do not conflict with the Amish lifestyle.

Opponents of genetically modified food often refer to it as "Frankenfood", after Mary Shelley's character Frankenstein and the monster he creates, in her novel of the same name. The term was coined in 1992 by Paul Lewis, an English professor at Boston College who used the word in a letter he wrote to the *New York Times* in response to the decision of the US Food and Drug Administration to allow companies to market genetically modified food. The term "Frankenfood" has become a battle cry of the European side in the US-EU agricultural trade war.

The authors of *The Frankenfood Myth* provide some support for genetically modified food:

- Henry I. Miller of Stanford's Hoover Institution and Gregory Conko of the Competitive Enterprise Institute make the case that foods modified by recombinant DNA splicing present no new or special dangers, but in fact may improve the lives of countless millions worldwide.

International trade of genetically modified foods

The European Union and the United States have strong disagreements over the EU's regulation of genetically modified food. The US claims these regulations violate free trade agreements; the EU counter-position is that free trade is not truly free without informed consent.

In Europe, a series of unrelated food crises during the 1990s created consumer apprehension about food safety in general, eroded public trust in government oversight of the food industry, and left some consumers unwilling to consider "science" to be a guarantee of quality. This has further fueled widespread public concern about genetically modified organisms (GMO), in terms of potential environmental protection (in particular biodiversity), health, and safety of consumers. Critics of GM foods contend that there is evidence that the cultivation of genetically modified plants may lead to environmental changes. Directives such as directive 2001/18/EC were designed to require authorisation for the placing GMO on the market, in accordance with the precautionary principle.

Many European consumers are demanding the right to make an informed choice about whether or not to consume GMO foods. Some polls indicate that Americans would also like labeling, but it has not yet become a major issue. New EU regulations should require strict labeling and traceability of all food and animal feed containing more than 0.5 percent GM ingredients. A 2003 survey by the Pew Research Centre found that a majority of people in all countries surveyed felt that GM foods were "bad". The lowest scores were in the US and Canada, where 55% and 63% (respectively) were against it, while the highest were in Germany and France with 81% and 89% disapproving. The survey also showed a strong tendency for women to be more opposed to GM foods than men.

In 2002, Oregon Ballot measures gave voters in that state one of the first opportunities in the United States to directly address that issue. The measure, which would have required the labeling of genetically engineered foods, failed to pass by a ratio of 7 to 3. Friedrich-Wilhelm Graefe zu Baringdorf, member of the German Green Party and vice president of the Landwirtschaftsausschuss (committee of agriculture) of the European Commission said on the 1 July 2003: "In America 55% of the consumers are against GM food and 90% in favour of a clear labeling."

The European Union and United States are in strong disagreement over the EU's ban on most genetically modified foods. The value of agricultural trade between the US and the European is estimated at $57 billion at the beginning of the 21st Century, and some in the U.S., especially farmers and food manufacturers, are concerned that the new proposal by the European Union could be a barrier to much of that trade. In 1998, the United States exported $63 million worth of maize to the EU, but the exports decreased to $12.5 million in 2002.

The drop-off might also be due to falling commodities prices, less demand due to the recession, U.S. maize being priced out of foreign markets by a strong dollar, and importing countries' reaction to the planned invasion of Iraq. Similar European public opposition to Israeli treatment of Palestinians has also affected Israeli food exports. However, American farm industry advocates blame the EU's ban.

The European Parliament's Committee on the Environment, Public Health and Food Safety proposal, adopted in the summer of 2002 and expected to be implemented in 2003, has deep cultural roots that are difficult to understand for the US agricultural community. It requires that all food/feed containing or derived from genetically modified organisms be labeled and any GM ingredients in food be traced. It would also require documentation tracing biotechnological products through each step of the grain handling and food production processes.

The new European tax, tariff and trade proposal would particularly affect US maize gluten and soybean exports, as a high percentage of these crops are genetically modified in the USA (about 25 percent of US maize and 65 percent of soybeans are genetically modified in 2002). The ultimate resolution of this case is widely thought to rest on labeling rather than food aid. Many European consumers are asking for food regulation (demanding labels that identify which food has been genetically modified), while the American agricultural industry is arguing for free trade and is strongly opposed to labeling, saying it gives the food a negative connotation.

Lori Wallach, director of Public Citizen's Global Watch indicates that American agricultural industry is "using trade agreements to determine domestic health, safety and environmental rules" because they fear that "by starting to distinguish which food is genetically modified, then they will have to distinguish energy standards, toxic standards that are different to those that European promotes."

The American Agricultural Department officials answer that since the United States does not require labeling, Europe should not require labeling either. They claim mandatory labelling could imply there is something wrong with genetically modified food, which would be also a trade barrier. Current U.S. laws do not require GM crops to be labeled or traced because U.S. regulators do not believe that GM crops pose any unique risks over conventional food. Europe answers that the labeling and traceability requirements are not only limited to GM food, but will apply to any agricultural goods. The American agricultural industry also complains about the costs implied by labeling.

The ban over agricultural biotechnology things is said by some Americans to breach World Trade Organisation rules. Robert B. Zoellick, the United States trade representative, indicated the European position toward GMO was thought of as "immoral" since it could lead to starvation in the [{developing country/developing world} of wars], as seen in some famine-threatened African countries (eg, Zambia, Zimbabwe, and Mozambique) that refuse to accept US aid because it contains GM food.

Zoellick's critics argue that US concern over Third World starvation is merely a cover for other issues. Some money for development aid is used by the American government via the World Food Programme (WFP) to help their farmers by buying up overproduction and giving it to the UN organisation. GM-scepticism interferes with this programme.

American farmers lost marketshare in certain countries after changing to genetically modified food because of sceptical consumers. Another European response to the claims of immorality is that the EU gives 7 times more in development aid than the US, yet its economy is less than 10% bigger than America's, and its GDP/head much lower than that of the US.

In May 2003, after initial delay due to the war against Iraq, the Bush administration officially accused the European Union of violating international trade agreements, in blocking imports of U.S. farm products through its long-standing ban on genetically modified food. Robert Zoellick announced the filing of a formal complaint with the WTO challenging the moratorium after months of negotiations trying to get it lifted voluntarily. The complaint was also filed by Argentina, Canada, Egypt, Australia, New Zealand, Mexico, Chile, Colombia, El Salvador, Honduras, Peru, and Uruguay. The formal WTO case challenging the EU's regulatory system was in particular lobbied by U.S. biotechnology giants like Monsanto and Aventis and big agricultural groups such as the National Corn Growers Association.

EU officials questioned the action, saying it will further damage trade relations already strained by the U.S. decision to launch a war against Iraq despite opposition from members of the U.N. Security Council. The US move was also interpreted as a sanction against the EU for requesting the end of illegal tax breaks for exporters or face up to $4 billion in trade sanctions in retaliation for Washington's failure to change the tax law, which the WTO ruled illegal four years ago.

Ratification of the Biosafety Protocol

In June 2003, the European Parliament ratified a three-year-old U.N. biosafety protocol regulating international trade in genetically modified food, expected to come into force in fall 2003 since the necessary number of ratification was reached in May 2003. The protocol lets countries ban imports of a genetically modified product if they feel there is not enough scientific evidence the product is safe and requires exporters to label shipments containing genetically altered commodities such as corn or cotton. It makes clear that products from new technologies must be based on the precautionary principle and allows developing nations to balance public health against economic benefits.

Jonas Sjoestedt, a Swedish Left member of the EU assembly, said that "this legislation should help the EU to counter recent accusations by the U.S administration that the EU is to blame for the African rejection of GM food aid last year." The United States did not sign the protocol, saying it was opposed to labeling and fought import bans. In 1998, a *de facto* moratorium led to the suspension of approvals of new genetically modified organisms (GMO) in the European Union pending the adoption of revised rules to govern the approval, marketing and labelling of biotech products. Imports and cultivation of already approved GM varieties and food products continued. In July 2003, European environment ministers and the European Parliament agreed to new controls on GMOs that could eventually lead the then 15 members bloc to re-open the Union's markets to new genetically modified products in 2004.

The new labeling and traceability rules, which cover both food and feed, require any products with a GMO content of more than 0.9 percent to be labelled. Labelling is also required for products that have been derived from GMOs, but where the GM content might no longer be detectable (such as soy oil produced from genetically modified soy). The threshold for the presence of unapproved GMOs is 0.5 percent provided that the GMOs have been judged as safe for human health and the environment by the relevant Scientific Committees or the European Food Authority. This amount will be set for 3 years. After 3 years, all food containing non-authorized GMO will be banned. Animals fed with transgenic cereals are not covered by the labeling requirements.

Traceability of GMO products is mandatory, from sowing to final product. Genetically modified goods will have to carry a special harmless DNA sequence (a DNA code bar) identifying the origin of the crops, making it easier for regulators to spot contaminated crops, feed, or food, and enabling products to be withdrawn from the food chain should problems arise. A series of additional sequences of DNA with encrypted information about the company or what was done to the product could also be added to provide more data.

Following the entry into force of the new regulations, the first genetically modified food product (canned maize) since 1998 was approved for marketing in the European Union in May 2004. While a number of other biotech products have been approved since then, approvals remain controversial. European ministers have continuously failed to reach a decision in support of or against the applications, highlighting the big divide among member states. As a result, the approvals were granted by the European Commission, which is entitled to take a decision in case ministers fail to do so.

The U.S. population has historically placed a considerable degree of trust in the regulatory oversight provided by the U.S. Department of Agriculture and its agencies. There is little tradition of people having a close relationship with their food, with the overwhelming majority of people having bought their food in supermarkets for years. But the 2003 survey by the Pew Research Centre showed that even in the U.S., 55% see GM food as "bad" food.

In Europe, and particularly in the UK, there is less trust of regulatory oversight of the food chain. In many parts of Europe, a larger measure of food is produced by small, local growers using traditional (non-intensive & organic) methods.

Arpad Pusztai, considered by many to be the leading expert on GM foods, was silenced with threats of a lawsuit after he unexpectedly discovered that rats fed an experimental GM food developed immune system damage and other serious health problems in just ten days. Pusztai later reviewed an industry-sponsored study and found that seven of forty rats fed a GM crop died within two weeks; others developed stomach lesions. The crop was approved without further tests.

In May, when the U.S. filed a challenge with the World Trade Organization (WTO) disputing Europe's GM food policy, Trade Representative Robert Zoellick stated, "Overwhelming scientific research shows that biotech foods are safe and healthy." According to Andrew Kimbrell, director of the Centre for Food Safety, "The evidence in the book Seeds of Deception refutes U.S. science and safety claims, and undermines the basis of their WTO challenge."

Kimbrell says, "Author Jeffrey M. Smith's book also presents a compelling argument that nations may use to ban GM foods altogether." Countries gained the right to impose such a ban on September 11, three months after the UN biosafety protocol was signed by 50 nations. "The revelations in the book," says Kimbrell, "are being made public at a pivotal time in the global GM debate, and could tip the scales against the biotech industry."

The World Trade Organization has made a preliminary ruling that European Union restrictions on genetically engineered crops violate international trade rules. The United States, Canada, and Argentina together grow 80 percent of all biotech crops sold commercially, by which the EU regulates such crops. The countries argued that the EU's regulatory process was far too slow and its standards were unreasonable given that the overwhelming body of scientific evidence finds the crops safe.

8

Genetically Modified Crops

All living organisms have the ability to improve themselves through natural means in order to adapt to changing environmental conditions. However, it takes hundreds of years before any detectable improvement is obtained. Man then learned how to domesticate and breed plants in order to develop crops to his own liking and needs using various means including biotechnology.

Biotechnology is defined as a set of tools that uses living organisms (or parts of organisms) to make or modify a product, improve plants, trees or animals, or develop microorganisms for specific uses. Agricultural biotechnology is the term used in crop and livestock improvement through biotechnology tools. This monograph will focus only on agricultural crop biotechnology. Biotechnology encompasses a number of tools and elements of conventional breeding techniques, bioinformatics, microbiology, molecular genetics, biochemistry, plant physiology, and molecular biology.

The biotechnology tools that are important for agricultural biotechnology include:

- Conventional plant breeding
- Tissue culture and micropropagation
- Molecular breeding or marker assisted selection
- Genetic engineering and GM crops
- Molecular Diagnostic Tools

Conventional Plant Breeding

Since the beginning of agriculture eight to ten thousand years ago, farmers have been altering the genetic makeup of the crops they grow. Early farmers selected the best looking plants and seeds and saved them to plant for the next year. The selection for features such as faster growth, higher yields, pest and disease resistance, larger seeds, or sweeter fruits has dramatically changed domesticated plant species compared to their wild relatives. Plant breeding came into being when man learned that crop plants could be artificially mated or crosspollinated to be able to improve the characters of the plant. Desirable characteristics from different parent plants could be combined in the offspring. When the science of plant breeding was further developed in the 20th century, plant breeders understood better how to select superior plants and breed them to create new and improved varieties of different crops.

This has dramatically increased the productivity and quality of the plants we grow for food, feed and fibre. Conventional plant breeding has been the method used to develop new varieties of crops for hundreds of years. However, conventional plant breeding can no longer sustain the global demand with the increasing population, decline in agricultural resources such as land and water, and the apparent plateauing of the yield curve of the staple crops. Thus, new crop improvement technologies should be developed and utilized.

Fig. Conventional Breeding Entails Sexual Hybridization Followed by Careful Selection

Mutation Breeding

The art of recognizing desirable traits and incorporating them into future generations is very important in plant breeding. Breeders inspect their fields and travel long distances in search of individual plants that exhibit desirable traits. A few of these traits occasionally arise spontaneously through a process called mutation, but the natural rate of mutation is very slow and unreliable to produce plants that breeders would like to see. In the late 1920s, researchers discovered that they could greatly increase the number of these variations or mutations by exposing plants to X-rays and mutation-inducing chemicals.

"Mutation breeding" accelerated after World War II, when the techniques of the nuclear age became widely available. Plants were exposed to gamma rays, protons, neutrons, alpha particles, and beta particles to see if these would induce useful mutations. Chemicals such as sodium azide and ethyl methanesulphonate, were also used to cause mutations. Mutation breeding efforts continue around the world today. Of the 2,252 officially released mutation-derived varieties, 1,019 or almost half have been released during the last 15 years. Some varieties of wheat, barley, rice, potatoes, soybeans, onions and others were produced via mutation breeding with agronomically-desirable characteristics.

Pure Line and Hybrid Seed Technology

The end result of plant breeding is either an open-pollinated (OP for corn) or inbred (for rice) varieties or an F1 (first filial generation) hybrid variety. OP and inbred varieties, when maintained and properly selected and produced, retain the same characteristics when multiplied. Hybrid seeds are an improvement over OP and inbred seeds in terms of yield, resistance to pests and diseases, and time to maturity. Hybrid seeds are developed by the hybridization or crossing of diversely related parent lines.

Pure lines are off-springs of several cycles of repeated self-pollination that "breed true" or produce sexual offspring that closely resemble their parents. Pure line development involves firstly, the selection of lines in the existing germplasm which express the desired characteristics such as resistance to pest and diseases, early maturity, yield, and others. These traits may not be present in only one line, thus selected lines are bred together by hand. In self-pollinated plants, flowers are emasculated by removing the anthers or the male part of the flower by hand, and are pollinated by pollen from another line. The female parent is usually the line that possesses the desired agronomic trait while the male parent is the donor of the new trait.

F1 (first filial generation) off-springs are planted and selfed, as well as the F2 generation. Breeders then select in the F3 and F4 generation the lines which exhibit their desired agronomic characteristics and the added trait. Testing for resistances to pests and abiotic stresses are conducted also at this time. Lines with desired traits and are rated intermediate to resistant/tolerant to the pests and abiotic stresses are selected and selfed in two to three more generations.

Lines which do not lose the new traits and are stable are termed pure lines and are stable. In hybrid seed technology, two pure lines with complementing traits and are derived from diversely related parents are bred together by hand. F1 hybrids are tested for hybrid vigour in all agronomic and yield parameters and compared to both parents. The resulting off-springs will usually perform more vigorously than either parents. Since the technology has been developed, it has brought tremendous impact in major crops including rice, corn, wheat, cotton, and other crops including many vegetables.

In the USA, the widespread use of corn hybrids, coupled with improved cultural practices by farmers, has more than tripled corn grain yields over the past 50 years from an average of 35 bushels per acre in the 1930s to 115 bushels per acre in the 1990s. No other major crop anywhere in the world even comes close to equaling that sort of success story. Hybrid rice technology helped China to increase its rice production from 140 million tons in 1978 to 188 million tons in 1990. Research at the International Rice Research Institute (IRRI) and in other countries indicates that hybrid rice technology offers opportunities for increasing rice yields by 15-20% beyond those achievable with improved, semi-dwarf, inbred varieties. With the proven impact of hybrid seed technology, new tools for hybrid breeding were discovered and utilized for self-pollinating crops including cytoplasmic male sterility (cms). Cytoplasmic male sterility is a condition where the plant is unable to produce functional pollen and would rely on other pollen source to produce seeds. This greatly facilitates large scale hybrid seed production, by-passing hand pollination.

Current hybrid seed technology uses three lines in order to produce the hybrid seed:

- The A line which contains a defective mitochondrial genome in the cytoplasm and a suppressed restorer gene,
- The B line which is genetically similar to the A line but contains a normal cytoplasm and a suppressed restorer gene,
- The restorer line, a distinctly unrelated line which contains normal cytoplasm and an active restorer gene (dominant).

The two line hybrid system, another hybrid seed technology relies on temperature and geographic location affecting the nuclear genome of the plant, manifested as male sterile. Hybrid seed technology assures hybrid vigour in the progenies but discovery and development of cms lines requires a lot of work and time.

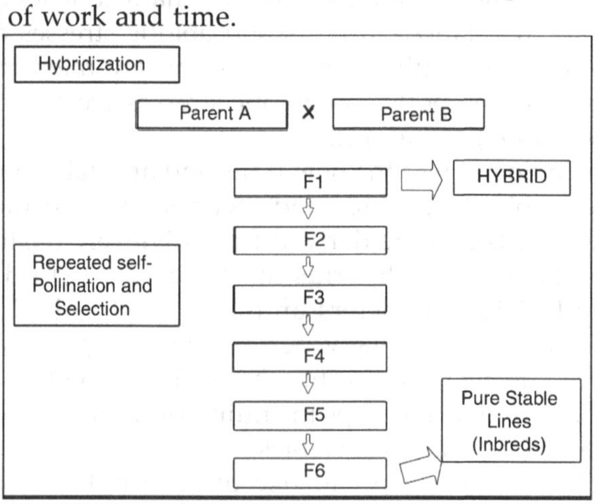

Fig. Pure Line (Inbred Line) Development

Conventional plant breeding resulting in open pollinated varieties or hybrid varieties has had a tremendous impact on agricultural productivity over the last decades. While an extremely important tool, conventional plant breeding also has its limitations. First, breeding can only be done between two plants that can sexually mate with each other. This limits the new traits that can be added to those that already exist in that species. Second, when plants are crossed, many traits are transferred along with the trait of interest including traits with undesirable effects on yield potential. Agricultural biotechnology is an option for breeders to overcome these problems.

Tissue Culture and Micropropagation

Plants usually reproduce through sexual means—they have flowers and seeds to create the next generation. Egg cells in the flowers are fertilized by pollen from the stamens (male part) of the flower of the same plant (self-pollination) or another plant (cross). Each of these sexual cells contains genetic material in the form of DNA. During sexual reproduction, DNA from both parents is combined creating offsprings similar to the parents (in self-pollinated crops), or in new and unpredictable ways, creating unique organisms (in cross-pollinated crops). Some plants and trees on the other hand need several years before they flower and set seeds, making plant improvement difficult.

Plant scientists have developed the science and art of tissue culture to assist breeders in this task. Tissue culture is the cultivation of plant cells, tissues, or organs on specially formulated nutrient media. Under the right conditions, an entire plant can be regenerated from a single cell. Plant tissue culture is a technique that has been around for more than 30 years. There are several types of tissue culture depending on the part of the plant (explant) used.

Anther Culture

Anther culture is a tissue culture method used to develop improved varieties in a short time. Pollen within an anther contains half dose of the genome (haploid) which spontaneously double (diploid) during culture. In some species however, colchicine treatment is necessary to induce doubling. Doubling of the genome will allow the expression of recessive traits which were suppressed, masked or undetected in routine plant breeding. Anthers are placed in a special medium, and immature pollen within the anther divide and produce a mass of dividing cells termed as callus. Healthy calli (plural of callus) are picked and placed in another medium to produce shoots and roots (regeneration). Stable plantlets are allowed to grow and mature in the greenhouse. Plant breeders can then select the desired plants from among the regenerated plants. Anther culture of F1 plants which are progenies in a specific breeding objective would allow many more different types of regenerants.

This is because the genetic constitution of the pollen will be more varied than those from the inbreds, thus breeders will have a wider range of traits to choose from. This technology has been employed in the successful development of doubled haploid lines of rice, wheat, sorghum, barley, and other field crops.

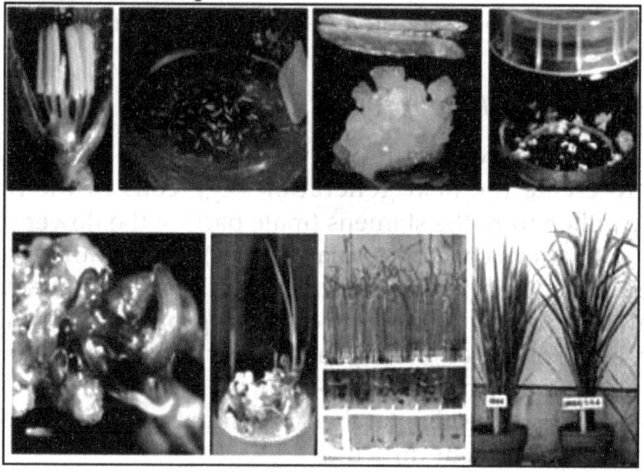

Fig. Anther Culture of Rice

Micropopagation

Micropopagation is a tissue culture method developed for the production of disease-free, high quality planting material and for rapid production of many uniform plants.

Actively-dividing young cells (meristem) are placed in a special medium and treated with plant hormones to produce many similar sister plantlets.

Since the meristem divides faster than disease-causing virus, clean materials are propagated and hundreds of uniform plantlets are produced in a short time.

Through micropropagation, it is now possible to provide clean and uniform planting materials in plantations—oil palm, plantain, pine, banana, abaca, date, rubber tree; field crops—eggplant, jojoba, pineapple, tomato; root crops—cassava, yam, sweet potato; and many ornamental plants such as orchids and anthuriums.

Micropropagated plants were found to establish more quickly, grow more vigorously and taller, have a shorter and more uniform production cycle, and produce higher yields than conventional propagules.

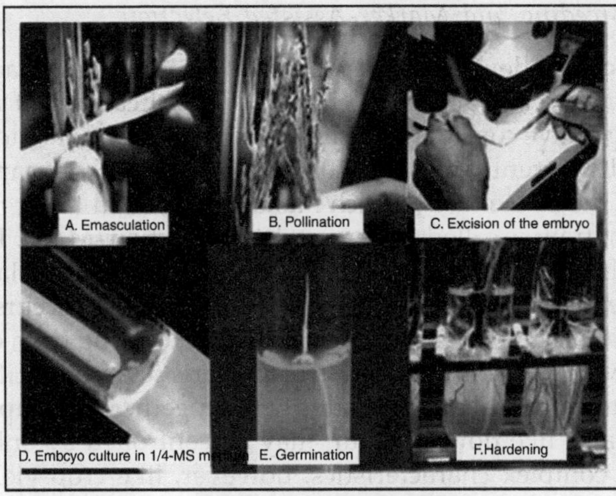

Fig. Embryo Rescue

Embryo Rescue

Embryo rescue involves the culture of immature embryos of plants in a special medium to prevent abortion of the young embryo and to support its germination. This is used routinely in breeding parental lines having different or incompatible genome such as in introducing important traits of wild relatives into cultivated crops. The development of a new rice plant type for West Africa (NERICA—New Rice for Africa) was a result of wide crosses between the Asian Oryza sativa and the African rice Oryza glaberrima.

It employs embryo rescue in the initial breeding and in the successive back crossing work followed by anther culture to stabilize the breeding lines. The new plants had combined yield traits of the sativa parent with local adaptation traits from glaberrima. Wild rices are a rich source of traits for resistance to pests and abiotic stresses. At the International Rice Research Institute, embryo rescue is utilized and facilitated the transfer of bacterial blight resistance genes from wild rice Oryza longistaminata to variety IR24 resulting to a bacterial blight resistant line (IRBB21).

Oryza rufipogon is a source of tungro resistance to a number of rice varieties. Plant tissue culture belongs to the lower end of the agricultural biotechnology ladder. But the plant's ability to regenerate a new plant is an important requisite in the development of improved crops through agricultural biotechnology. Plant tissue culture is a straightforward technique and many developing countries have already mastered it. Its application only requires a sterile workplace, nursery, and green house, and trained manpower. Unfortunately, tissue culture is labour intensive, time consuming, and can be costly.

Molecular Breeding and Marker-Assisted Selection

The process of developing new crop varieties requires many steps and can take almost 25 years. Now, however, applications of agricultural biotechnology have considerably shortened the time it takes to bring them to market. It currently takes 7-10 years for new crop varieties to be developed. One of the tools, which make it easier and faster for scientists to select plant traits is called marker-assisted selection (MAS). The different traits and physical features of plants are encoded in the plant's genetic material, the deoxyribonucleic acid (DNA). The DNA occurs in pairs of chromosomes (strands of genetic material), one coming from each parent. The genes, which control the plant's characteristics, are specific segments of each chromosome. All of the plant's genes together make up its genome.

Some traits, like flower colour, may be controlled by only one gene. Other more complex characteristics, however, like crop yield or starch content, maybe influenced by many genes. Traditionally, plant breeders have selected plants based on their visible or measurable traits, called the phenotype. But, this process can be difficult, slow, influenced by the environment, and costly—not only in the development itself, but also for the economy, as farmers suffer crop losses. As a shortcut, plant breeders now use molecular marker-assisted selection.

To help identify specific genes, scientists use what are called molecular markers which are short strings or sequence of nucleic acid which makes up a segment of DNA. The markers are located near the DNA sequence of the desired gene. Since the markers and the genes are close together on the same chromosome, they tend to stay together as each generation of plants is produced.

This is called genetic linkage. This linkage helps scientists to predict whether a plant will have the desired gene. If researchers can find the marker for the gene, it means the gene itself is present. As scientists learn where each of the markers occurs on a chromosome, and how close it is to a specific gene, they can create a map of the markers and genes on specific chromosomes. This genetic linkage map shows the location of markers and genes, and their distance from other known genes. Scientists can produce detailed maps in only one generation of plant breeding. Previously, scientists produced very simple genetic maps using conventional techniques.

It was observed long ago that as generations of plants were crossed, some traits consistently appeared together in the new generations (genetic linkage). However, since researchers could concentrate on only a few traits in each attempt at cross-breeding, it took many crosses to obtain even a very simple genetic map. Using very detailed genetic maps and better knowledge of the molecular structure of a plant's DNA, researchers can analyse a tiny bit of tissue from a newly germinated seedling. They don't have to wait for the seedling to grow into a mature plant to test for the presence of the specific trait. Once the tissue is analysed through molecular techniques, scientists know whether that seedling contains the appropriate gene.

If it doesn't, they can quickly move on and concentrate analysis on another seedling, eventually working only with the plants which contain the specific trait. Currently, molecular marker-assisted breeding, an agricultural biotechnology tool is already a routine step in breeding of most crops where the gene and the markers for a specific trait are known. This technique is being used in the efficient introgression of important genes into rice such as bacterial blight resistance, increased beta carotene content, and submergence tolerance to name a few.

Molecular markers are also used to determine the genetic profile of a line or variety. Random primers are used to scan the genomic constitution of the plant through molecular methods. The information is fed to a computer programme that will analyse the relatedness of one line to another. The information on genetic diversity of the lines is utilized in selecting for extremely unrelated parents useful for hybrid seed technology. The information will also provide details on the parentage of the line, the possible traits, and the unique identity of the plant useful for germplasm collection database.

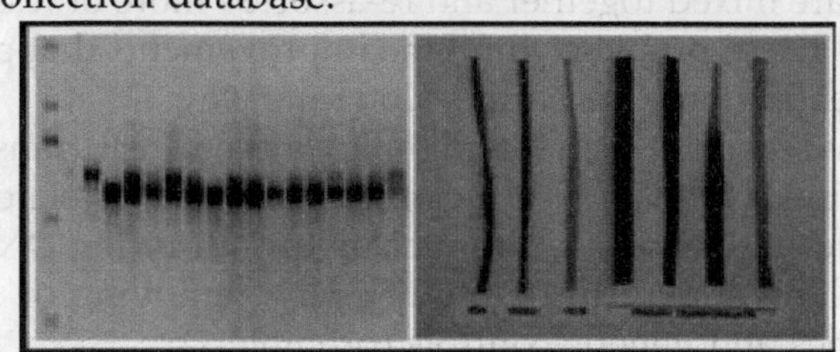

Fig. Molecular Marker-assisted Breeding

Increasing selection efficiency by selecting for markers associated/ linked with the trait of interest It should be noted, however, that molecular breeding through marker assisted selection is somewhat limited in scope compared to genetic engineering or modification because:

- Tt only works for traits already present in a crop;
- Tt cannot be used effectively to breed crops which have long generation time (*e.g.* citrus);
- Tt cannot be used effectively with crops which are clonally propagated because they are sterile or their offsprings does not resemble the parents.

This includes many staples such as yams, bananas, plantain, sweet potato, and cassava.

Genetic Engineering and GM Crops

Over the last 30 years, the field of agricultural biotechnology has developed rapidly due to the greater understanding of DNA as the chemical double-helix code from which genes are made. Genetic engineering is one of the modern agricultural biotechnology tools that is based on recombinant DNA technology.

The term genetic engineering, often interchanged with terms such as gene technology, genetic modification, or gene manipulation, is used to describe the process by which the genetic makeup of an organism can be altered using "recombinant DNA technology."

This involves using laboratory tools and specific enzymes to cut out, insert, and alter pieces of DNA that contain one or more genes of interest. The ability to manipulate individual genes and to transfer genes between species that would not readily interbreed is what distinguishes genetic engineering from traditional plant breeding. With conventional plant breeding, there is little or no guarantee of obtaining any particular gene combination from the millions of crosses generated.

Undesirable genes can be transferred along with desirable genes or while one desirable gene is gained, another is lost because the genes of both parents are mixed together and re-assorted more or less randomly in the offspring. These problems limit the improvements that plant breeders can achieve, eating time and funds along the way.

In contrast, genetic engineering allows the direct transfer of one or just a few genes, between either closely or distantly related organisms. Not all genetic engineering techniques involve inserting DNA from other organisms. Plants may also be modified by removing or switching off particular genes and genetic controls (promoters).

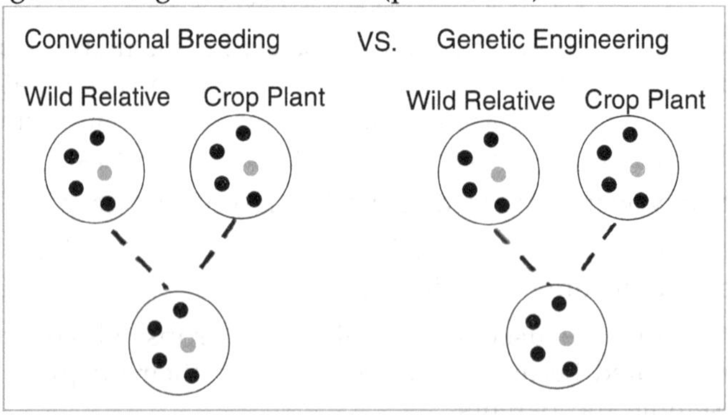

Fig. Conventional vs. Genetic Engineering

Application of Genetic Engineering in Crop Production

Genetic engineering techniques are only used when all other techniques have been exhausted and when:

- The trait to be introduced is not present in the germplasm of the crop;
- The trait is very difficult to improve by conventional breeding methods;
- Tt will take a very long time to introduce and/or improve such trait in the crop by conventional breeding methods.

Modern plant breeding is a multi-disciplinary and coordinated process where a large number of tools and elements of conventional breeding techniques, bioinformatics, biochemistry, molecular genetics, molecular biology and genetic engineering are utilized and integrated.

Development of Transgenic Crops

Although there are many diverse and complex techniques involved in genetic engineering, its basic principles are reasonably simple. It is however, very important to know the biochemical and physiological mechanisms of action, regulation of gene expression and safety of gene and gene product to be utilized. The process of genetic engineering requires the successful completion of a series of six steps.

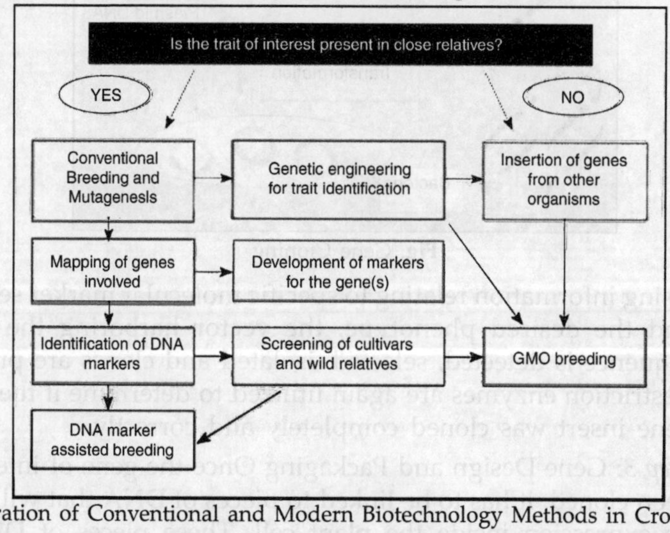

Fig. Integration of Conventional and Modern Biotechnology Methods in Crop Breeding

- *Step 1*: Nucleic acid (DNA/RNA) Extraction Nucleic acid extraction, either DNA or ribonucleic acid (RNA) is the first step in the genetic engineering process. It is therefore important that reliable methods are available for isolating these components from the cell. In any isolation procedure, the initial step is the disruption of the cell of the desired organism, which may be viral, bacterial or plant cells, in order to extract the nucleic acid. After a series of chemical and biochemical steps, the extracted nucleic acid can be precipitated to form thread-like pellets of DNA/RNA.

- *Step 2*: Gene cloning The second step is gene cloning. There are basically four stages in any cloning experiment: generation of DNA fragments, joining to a vector, propagation in a host cell, and selection of the required sequence. In DNA extraction, all DNA from the desired organism is extracted. This genomic DNA is treated with specific enzymes called restriction enzymes cutting it into smaller fragments with defined ends to allow it to be cloned into bacterial vectors. Copies of the vector will then harbor many different inserts of the genome. These vectors are transformed into bacterial cells and thousands of copies are produced.

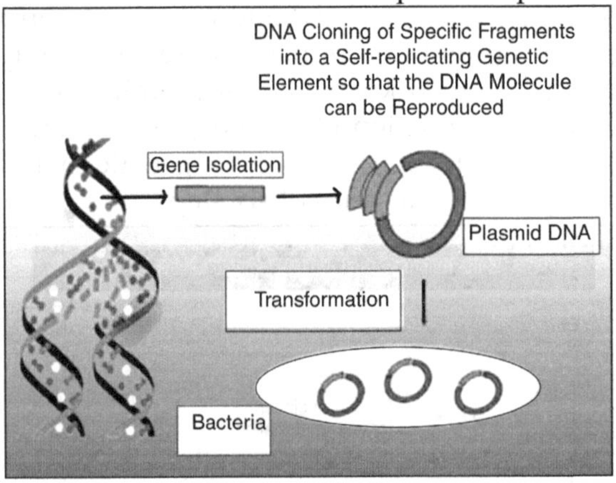

Fig. Gene Cloning

- Using information relating to specific molecular marker sequences and the desired phenotype, the vector harboring the desired sequence is detected, selected, isolated and clones are produced. Restriction enzymes are again utilized to determine if the desired gene insert was cloned completely and correctly.
- *Step 3*: Gene Design and Packaging Once the gene of interest has been cloned, it has to be linked to pieces of DNA that will control its expression inside the plant cell. These pieces of DNA will switch on (promoter) and off (terminator) the expression of the gene inserted. Gene designing/packaging can be done by replacing an existing promoter with a new one, incorporating a selectable

marker gene and reporter gene, adding gene enhancer fragments, introns, and organelle-localizing sequences, among others. Promoters Promoters allow differential expression of genes. For instance some promoters cause the inserted genes to be expressed all the time, in all parts of the plant (constitutive) whereas others allow expression only at certain stages of plant growth, in certain plant tissues, or in response to external environmental signals. The amount of the gene product to be expressed is also controlled by the promoter. Some promoters are weak, whereas others are strong. Controlling the gene expression is an advantage in developing GM plants.

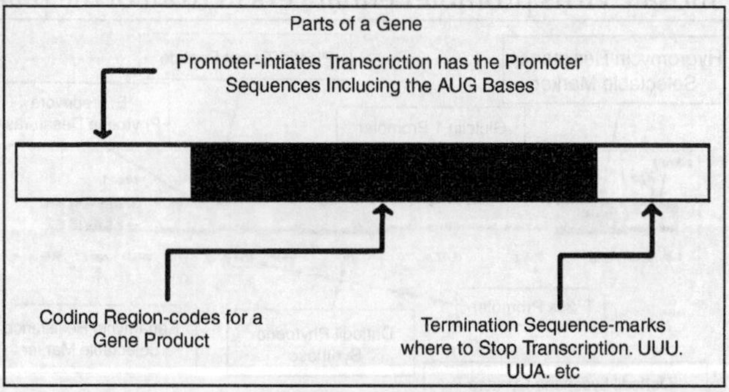

Fig. Parts of a Gene

- *Marker Genes*: Selectable marker genes are usually linked to the gene of interest to facilitate its detection once inside the plant tissues. This enables the selection of cells that have been successfully incorporated with the gene of interest, thus saving considerable expense and effort. Genetic engineers used antibiotic resistance and herbicide resistance marker genes to detect cells that contain the inserted gene. Cells that survive the addition of marker agents to the growth medium indicate the presence of the inserted gene. Although increase in antibiotic resistance in humans and animals is unlikely to occur using antibiotic resistance marker, genes coding for resistance to non-medically important antibiotics are preferred. In addition, alternative types of marker genes have been developed which are related to plant metabolism such as phosphomannose isomerase, xylose isomerase and others.
- *Reporter Genes*: Reporter genes are cloned into the vector in close proximity to the gene of interest, to facilitate the identification of transformed cells as well as to determine the correct expression of the inserted gene. Reporter genes that

have been used include: the beta glucuronidase gene (gusA gene) which acts on a particular substrate producing a blue product, hence making the transformed cells blue; the green fluorescent protein (gfp) which allows transformed cells to glow under a green light; and luciferase gene that allows cells to glow in the dark, among others.

- *Enhancers*: Several genetic sequences can also be cloned in front of the promoter sequences (enhancers) or within the genetic sequence itself (introns, or non coding sequences) to promote gene expression. An example is the cloning of the cauliflower mosaic virus promoter enhancers in front of the plant promoter.

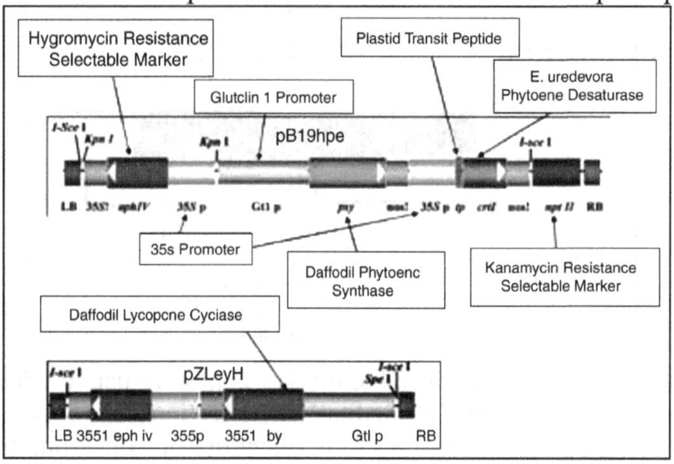

Fig. Components of a Gene Construct Used in Developing Golden Rice

- Once the gene of interest is packaged together (with the promoter, reporter and the marker gene), it is then introduced into a bacterium to allow for the creation of many copies of the gene package. The DNA isolated from the bacterial clones can then be used for plant cell transformation using particle bombardment. If however the use of bacteria Agrobacterium tumefaciens is preferred in the plant transformation, the whole gene package should be cloned in between two border sequences (left and right border) of a binary vector.
- This will allow processing of the Agrobacterium so that only the transfer DNA (T-DNA) will be incorporated into the plant genome:

- *Step 4.* Transformation The most common methods used to introduce the gene package into the plant cells in a process called transformation or gene insertion, include biolistic transformation using the gene gun and Agrobacterium-mediated transformation.

– *Particle Bombardment*: Particle bombardment is a mechanical method of introducing the desired gene. The desired genetic sequence is cloned into a plant DNA vector and introduced into the plant using the gene gun or particle gun. As in the common gun, the gene gun uses minute particles of tungsten or gold as the bullet. These particles are coated with the DNA solution and fired to the plant cells through the force of the Helium gas inside a vacuum-filled chamber. The DNA and the tungsten/ gold particles get inside the cell, and within 12 hours, the inserted DNA gets inside the nucleus and integrated with the plant DNA. The tungsten/ gold particles are sequestered to the vacuole and eliminated later.

Transformed cells are cultured in vitro and induced to form small plants (regeneration) that express the inserted gene.

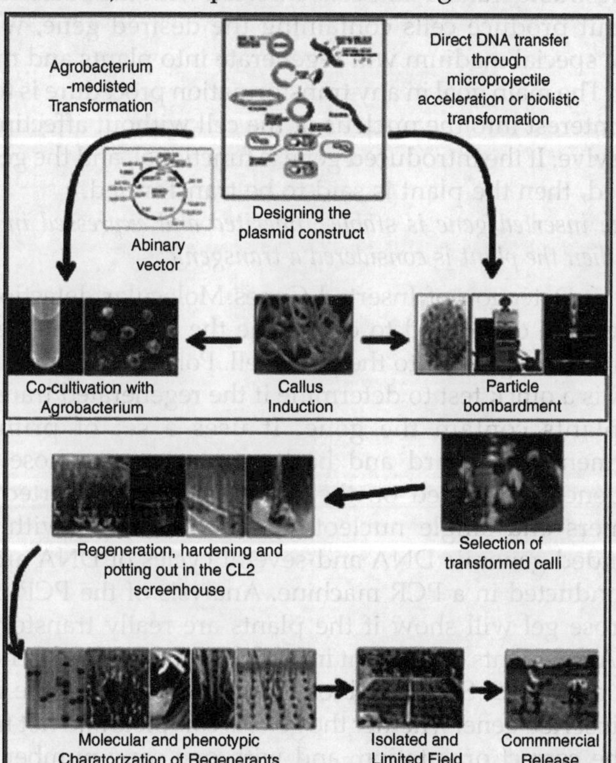

Fig. Genetic Transformation Methods (Biolistics or Gene Gun and Agrobacterium Tumefaciens-mediated Transformation Methods)

Agrobacterium Tumefaciens-meDiated Transformation

The "sharing" of DNA among living forms is well documented as a natural phenomenon. For thousands of years, genes have moved from one organism to another. For example, Agrobacterium tumefaciens, a soil bacterium known as 'nature's own genetic engineer', has the natural ability to genetically engineer plants. It causes crown gall disease in a wide range of broad-leaved plants, such as apple, pear, peach, cherry, almond, raspberry and roses.

The disease gains its name from the large tumor-like swellings (galls) that typically occur at the crown of the plant, just above soil level. Basically, the bacterium transfers part of its DNA to the plant, and this DNA integrates into the plant's genome, causing the production of tumors and associated changes in plant metabolism. Molecular biologists have utilized this biological mechanism to improve crops.

The genes that cause the galls are removed and replaced with genes coding for desirable traits. Plant cells infected with the bacterium will not form galls but produce cells containing the desired gene, which when cultured in a special medium will regenerate into plants and manifest the desired trait. The main goal in any transformation procedure is to introduce the gene of interest into the nucleus of the cell without affecting the cell's ability to survive. If the introduced gene is functional, and the gene product is synthesized, then the plant is said to be transformed.

Once the inserted gene is stable, inherited and expressed in subsequent generations, then the plant is considered a transgenic:

- *Step 5*: Detection of Inserted Genes Molecular detection methods have been developed to determine the integrity of the transgene (introduced gene) into the plant cell. Polymerase chain reaction or PCR is a quick test to determine if the regenerated transgenic cells or plants contain the gene. It uses a set of primers (DNA fragments)—forward and backward primers, whose nucleotide sequences are based on the sequence of the inserted gene. The primers and single nucleotides are incubated with the single stranded genomic DNA and several cycles of DNA amplification is conducted in a PCR machine. Analysis of the PCR products in agarose gel will show if the plants are really transformed when DNA fragments equivalent in size with the inserted gene is present and amplified. Southern blot analysis determines the integrity of the inserted gene: whether the gene is complete and not fragmented, at the correct orientation, and with one copy number. The DNA coding sequence is the probe binding to the single nitrocellulose paper. Autoradiography will reveal the transgenic status of the plant. Northern blot analysis determines whether the transcript or the messenger RNA (mRNA) of the introduced DNA is present and is correctly transcribed in the transgenic plant. The messenger RNA of the transgenic plants are isolated and processed to bind to the nitrocellulose membrane. Labeled DNA is used to bind to the mRNA and can be visualized through autoradiography. Western blot analysis or protein immuno blotting is an analytical technique used to detect whether the transgenic plants produce the specific protein product of the introduced gene. Protein samples are extracted from the transgenic plants, processed into denatured proteins and transferred to a nitrocellulose membrane. The protein is then probed or detected using the antibodies specific to the target protein.

- *Step 6*: Backcross Breeding (if needed) Genetic transformation is usually conducted in elite or commercial varieties which already possess the desired agronomic traits but lacks the important trait of the transgene. Thus, once successfully conducted, the genetically modified plant will be easily recommended for commercialization if it shows stability in several generations and upon successfully passing and fulfilling varietal registration requirements. However, some plant transformations may have been performed in plant varieties which are amenable to genetic transformation but are not important in the target country, or in a variety adapted only in the country where the transformation was conducted. There may also be sterility problems in the transgenic plant. In such cases, conventional plant breeding is performed where the transgenic plant becomes the pollen source in the breeding programme and the elite lines or commercial varieties as the recurrent parent. Backcross breeding enables the combination of the desired traits of the recurrent parent and the transgenic line in the offsprings. The length of time in developing transgenic plant depends upon the gene, crop species, available resources and regulatory approval. It varies from 6 to 15 years before a new transgenic plant or hybrid is ready for commercial release.

Commercially Available Crops Improved through Genetic Engineering

Transgenic crops which are planted commercially are herbicide tolerant soybean, maize, canola, cotton; insect resistant maize and cotton; and virus resistant squash and papaya. With genetic engineering, more than one trait can be incorporated into a plant and are called stacked traits. These are currently corn and cotton crops with both herbicide and insect tolerance traits. Transgenic crops with combined traits are also available commercially such as the herbicide tolerant and insect resistant maize and cotton.

New and Future Initiatives in Crop Genetic Engineering

To date, commercial GM crops have delivered benefits in crop production, but there are also a number of products in the pipeline which will make more direct contributions to food quality, clean environment, pharmaceutical production, and livestock feeds.

Examples of these products include: rice with higher levels of iron and beta carotene (an important micronutrient which is converted to vitamin A in the body); long life banana that ripens faster on the tree and can therefore be harvested earlier; maize with improved feed value; delayed ripening papaya; papaya ringspot virus resistant papaya; tomatoes with high levels of flavonols, which are powerful antioxidants; drought tolerant maize and wheat; maize with improved phosphorus availability; arsenic-tolerant plants; insect resistant eggplant and rice; edible vaccines from fruit and vegetables; low lignin trees for paper making among others.

Food Safety

Safety of Genetically–Engineered Food Assessed

Before GE foods and products made from GE crops are approved for use, they have undergone safety testing by the companies or institutions that developed them. Data were reviewed by government regulatory agencies and scientific reviewers based on internationallyaccepted protocols. Frequently, GE foods are also tested by outside groups and the results published in peer-reviewed journals. The process is comparable to safety assessments for pharmaceutical drugs and biomarkers; pharmaceutical companies provide safety data that are subsequently reviewed by the U.S. Food and Drug Administration (FDA) scientists. To date all GE products in the marketplace have undergone full reviews by regulatory agencies regarding safety and content relative to unmodified forms. Submitting the safety data is in the developer's best interests given the legal liabilities incurred should a problem with the food arise following market introductions.

When DNA is Eaten

DNA is chemically identical regardless of its source and is mostly degraded during industrial processing and in the digestive tract. Small fragments can be detected in certain body tissues, such as leukocytes, liver, and spleen. The daily human intake of DNA in food is estimated at 0.1-1g. Estimates of the total daily transgene DNA intake can be calculated assuming 50% of the diet is from GE foods and transgenes represent an estimated 0.0005% of total DNA in food, as 0.5-5ug/day.

In July 2007, the European Food Safety Authority released statements on the fate of genes and proteins in food and feed: "After ingestion, a rapid degradation into short DNA or peptide fragments is observed in the gastrointestinal tract of animals and humans" and "To date, a large number of experimental studies with livestock have shown that recombinant DNA fragments or proteins derived from GM plants have not been detected in tissues, fluids or edible products of farm animals"

Changes in the Nutritional Content of Genetically-Engineered Food

GE foods are tested in comparison with conventional counterparts in terms of the nutritional composition: levels of protein, carbohydrate, fat, vitamin, mineral, fibre, moisture, and phytochemicals, and analysed if the composition is substantially equivalent. GE crops and conventional crops should have been grown in comparable conditions to eliminate the effect of the environment in the nutritional composition.

There are also GE crops which are developed to change the nutritional profiles of the foods such as those with increased B-carotene, flavonoids, calcium, folate, and iron availability. According to US-FDA policy, GE foods with altered nutritional traits must be labeled to indicate nutritional differences; one example is VistiveTM, a low-linoleic oil from GE soybeans that can be used instead of trans fat-containing oils. Such crops should be tested for substantial equivalence to compounds unrelated to the introduced trait.

Bt Protein Affect Hsumans

Bt proteins are naturally occurring insecticides produced by the soil bacterium, Bacillus thuringiensis, used to control crop pests such as larvae of butterflies and moths, beetles, and mosquitoes since the 1920s. The crystalline, inactive insecticidal Bt proteins, form bodies inside the bacterium and become active when they are eaten by the target insect larva and cleaved.

The active peptides bind to specialized receptors in the midgut of the insect, creating holes in the gut membrane that cause contents to leak and kill the larvae. The precision of different Bt proteins for their targets resides in the specificity of their tight binding to companion receptors in the insect gut. In recent years, a variety of safety studies were conducted specifically on native Bt proteins to show that they do not have characteristics of food allergens or toxins. Data on CryIAb in maize and cotton and Cry1Ac in tomato, maize and cotton have been carefully reviewed by regulatory agencies in numerous countries, including the U.S., Canada, Japan, UK, EU, Russia, and South Africa. Mycotoxin are toxic and carcinogenic chemicals produced by fungi that gain entry into the holes produced by the larva in corn. The reduction of mycotoxin incidence in Bt corn results in a positive impact in the improvement of corn yield, human and animal health.

Genetically Engineered Foods Cause Food Asllergies

No food is 100% safe, be it conventional, GE, or organic. Allergies are present in the "big eight" which is composed of milk, eggs, fish, shellfish, tree nuts, soybeans, wheat, and peanuts. Since food safety testing conducted on GE foods focuses on the introduced gene and its protein product, it seems unlikely that allergenicity issues related to a commercialized GE food that has undergone strict government health regulatory scrutiny will be greater than that of conventional foods, created by classical breeding and mutation that have not undergone such scrutiny.

Viral Genetic Sequences Inserted in the Genetically Engineered Crops Create a Human Risk

One of the important concerns is the use of the virus-derived promoter which are Introduced sequences in the transgenic plants that regulates how much, where, and when the encoded protein is expressed. This includes the cauliflower mosaic virus 35S which was used in some commercial GE crops, eg. Bt 11, Bt 176, Mon 810 maize, and Roundup Ready soybean.

Speculations that the "35S promoter could affect the stomach and colonic lining and cause a growth factor effect with the unproven possibility of hastening cancer formation in those organs" were forwarded earlier without any scientific experimentation.

These speculations have been extensively rebutted by the scientific community because the 35S promoter can be found everywhere in nature. For instance, an estimated 14-25% of oilseed rape in the field is infected with CaMV; similar numbers have been estimated for cauliflower and cabbage.

Because of its prevalence in foods, humans have consumed CaMV and its promoters at high levels for decades with no observable effects. The presence of the CaMV promoter in GE plants does not in principle present a different situation. Additionally, DNA in food is rapidly broken down during digestion, giving it little time to interact with the stomach and colonic linings.

Antibiotic Resistance Genes in Genetically Engineered Foods Increase Antibiotic Resistance in Humans and Animal Intestinal Fsslora

To develop antibiotic resistance in microorganisms present in the human and animal digestive tract, there should be a functional transfer of the antibiotic resistance gene, its controlling elements, and its integration in the bacterial chromosome.

This is next to impossible, since during chewing, cells in food are broken down. In raw food, as the cells are destroyed, DNA is released and highly active enzymes in the saliva and in the plant start degrading the DNA. This process continues in the digestive tract where other enzymes further breaks down DNA and proteins. In humans, food remains in the stomach for approximately 2 hours, where the remaining DNA is fragmented into small pieces. The antibiotic resistance gene from GE maize was shown not to transfer to gut bacteria in chickens fed with GE maize.

To refrain from using the controversial antibiotic resistance or herbicide tolerance genes as selectable markers, new selection strategies for identifying engineered plants have been developed. These include genes such as phosphomannose and xylose isomerase that facilitate selection by giving transgenic cells a metabolic advantage over non transgenic cells, as well as other means to excise the marker genes in the commercial product.

Genetic Engineering be Used to Make Pharmaceuticals

Plant-derived pharmaceuticals and vaccines for common diseases such as hepatitis B, pneumonic and bubonic plague, as well as against allergy sufferers, asthma, seasonal allergies and atopic dermatitis have been developed since the early 1990s.

Plant vaccines have the advantage of being readily consumed with limited or no processing without the need for cold storage.

However, these GE crops may enter the food supply if not properly handled and monitored. In the USA, where such pharmaceutical crops are cultivated, government regulations are in place. APHIS which regulates the movement and field testing of GE plants requires special steps to prevent plants that produce drugs or industrial enzymes from contaminating food crops:

- Labeling, packaging, and segregating regulated plant materials;
- Reproductive isolation to prevent GE pollen from fertilizing conventional plants;
- Postharvest monitoring to remove volunteer plants;
- Proper disposal of the transgenic materials.

This regulation was further strengthened in 2005 to include the following:

- Exclude field growth without a permit;
- Include crop inspections seven times a year, twice after harvest;
- Increase field isolation distances;
- Use dedicated farm equipment.

The National Corn Growers Association proposed safeguards such as:

- Using plants that are male-sterile or that produce non GE pollen,
- Dedicated production systems that isolate pharma crops,
- Third party verification
- Grower training programs.

In Sept. 2002, the FDA released a guidance document that recommends multiple strategies to prevent pharma crops from contaminating human or animal feed.

This documents suggests that those who are growing drug-producing plants that cross pollinate, such as corn and canola, strengthen containment procedures by growing plants in geographical regions where little or none of those crops are grown for food.

Labeling of Genetically Engineered Foods is not Required by the FDA

Government policy on labeling has been developed differently in many countries. In the USA, the FDA's labeling policy for GE foods is the same as for conventional foods and it assures that consumers are given information about nutritional, health safety or food quality changes in the end product. FDA mandated labels are not used to provide information about the process by which the food is made.

If a GE food is significantly different from its conventional counterpart, the food must be labeled to indicate the difference. Instances where the nutritional profile changes are included, for example if the GE food is created using genetic information from a previously recognized allergenic source, such as peanut, soy, or wheat, or if the new proteins has characteristics of known allergens.

For example, oils made from GE soybeans and canola varieties with changes in fatty acid composition must be labeled; foods containing those oils must be labeled and companies producing that oil must use a new name.

For example, Monsanto is using the name Vistive TM, to market its low-linoleic acid product from GE soybean oils. If a food contains a new potentially allergy-causing introduced protein, the label must state that the product contains the allergen and name its source.

Organic Foods

Organic farming is a method of agricultural production that does not allow the use of synthetic pesticides, fertilizers or growth enhancers. Food grown under organic certification differ from conventionally produced food by the manner in which they are grown, handled, and processed, but an "organic" label does not guarantee the nature of the product, the food, or ingredient, only its production method.

The important factors for many people who consume organic foods relate to the perceptions that they are healthier, taste better, are better for the environment, have lower pesticide levels and fewer food additives, and are better for animal welfare. However, organic certification does not imply that foods produced using organic methods are more nutritious or safer than those produced without organic methods. Differences reported in nutrient composition between organically and conventionally produced foods are interesting but it is very difficult to control all variables that might affect nutritional quality and ensure that the observed variations are significant and reproducible.

In addition, there are many important nutrients for which no significant differences have been found. Much more research is needed to determine whether food products are reproducible and have a significant impact on human health. Strictly from a nutritional perspective not enough data exist at present to show nutritional benefits from conventionally or organically produced foods that favour consuming either for health benefits. However, if the goal is to promote healthy eating, it is more important for consumers to focus on eating a healthy, balanced diet, rich in fruits and vegetables, than focusing on foods that are produced by particular methods. Convincing epidemiological evidence shows that diets rich in fresh fruits and vegetables, regardless of the methods used to produce them, improve health and are associated with reduced frequency and severity of a number of health conditions.

Environmental Issues

Rnsect Resistance to Bt

Resistance of insects against synthetic insecticides and Bt toxins in sprays occur and this will be true for GE crops. To slow this development in GE crops, several strategies have been developed. First generation GE crops produced only one Bt toxin in each plant. Planting refuges of non-Bt crops near Bt crops in the field is the primary strategy of delaying insect resistance. This is based on the idea that insects feeding on plants in the refuge are not selected for resistance. Insect resistance to Bt toxins is recessive. The heterozygous offsprings produced when homozygous resistant insects mate with susceptible insects are killed by the Bt crops. This high-dose/refuge strategy creates plants that produce Bt toxin concentrations high enough to kill heterozygous insects, making resistance functionally recessive. Insect resistance to Bt toxins can thus be postponed substantially. Another approach is called the pyramid or stacking strategy that combines two or more toxins in a single plant, each with different modes of action. An example is Bollgard II cotton producing Cry1Ac and Cry2b, which targets the same pest in two different ways.

Other approaches to delaying insect development are:

- Mixing seeds of Bt and non-Bt varieties are under small scale experiments
- The use of inducible promoter to drive Bt gene expression only during insect attack.
- Use of modified toxins to kill resistant insects, as exemplified by the use of modified Bt toxin that will not be affected by the mutations in the midgut cadherins.

Cadherins promote toxin oligomerization of Cry1A protein which has alpha helix in the binding site. Modified Cry1A which does not contain the alpha helix are independent of the cadherins and can thus be effective with insects which has developed resistance due to mutated or silenced cadherins To date, the elapsed time before the first cases of field resistance of insects to Bt crops were reported has been longer than what was predicted under worst-case scenarios, suggesting that management strategies may have delayed resistance development.

Despite documented cases of resistance, Bt crops remain useful against most target pests in most regions. As insect resistance to Cry toxins currently deployed in Bt crops increases, other strategies to create GE crops resistant to insects are being developed.

Genetically Engineered Crops Cause Adverse Effects on Non Target Organisms

Effects on GE crops on non target organisms have been studied with focus on:

- Monarch butterflies and black swallow tails. USA Environmental Protection Agency have concluded based on two studies that Bt corn was not a significant factor in field deaths of monarch larvae, particularly relative to factors such as the widespread use of pesticides and destruction of the butterfly's winter habits.
- Non target soil microorganisms. Studies on four maize varieties with two different Bt proteins (Cry1Ab and Cry3Bb1) versus near isogenic non-Bt varieties reveal that although numbers and types

of microbes and enzyme activities differ from season to season among varieties, no statistically significant differences were seen in number of different microbes, enzyme activities, or pH. Similar results were found comparing Bt and non-Bt cotton, and no Cry2Ab protein was detected in the rhizosphere in the field grown with Bt cotton.

- Non-target arthropods. Studies on foliage-dwelling arthropods on Bt maize expressing Cry3Bb1 compared with those of conventional insecticide treated maize show that there is no adverse impacts on abundance of any non target arthropods. Insecticide treated arthropods however reduced the number of non target insects: ladybird beetles, lacewings, and damsel bugs.
- Microbes and non target water insects. Water sediments and surface water after labeling genomic DNA of GE Bt corn revealed that sediments had more DNA than surface water. In addition, the Cry1Ab protein was not detectable in both samples.

Genetically Engineered Crops Result in the Population Decline of Other Organisms

Population decline of other organisms has been an ongoing phenomenon since man learned how to domesticate corps. The introduction of modern agricultural technologies including new varieties; competition between local and introduced varieties led to a displacement of local varieties; and displacing local varieties eroded genetic variability of regional crop populations. Extensive plant breeding in the early 1960s to feed the tremendous increase in the population produced high-yielding varieties of major food crops, resulting in yield increases but also significant displacement of traditional varieties and a concomitant loss in genetic diversity, particularly landraces of cereals and legumes.

Recognition of this fact led to establishment of genebanks across the globe with focus on specific crops. One issue on diversity is the gene flow from GE crops to wild and weedy relatives which could render selective advantage of recipients in certain environments. Gene flow can also happen naturally in conventionally bred and commercialized crops.

This is addressed by the adoption of measures needed in cultivating GE crops near centers of origin depending on the nature of the trait and the frequency of its introduction into an ecosystem. Currently, studies on impact assessment of transgenes moving into wild relatives and the potential to change ecosystem dynamics are requested in environmental impact statements before any GE plant is released. It provides insights into the possible outcomes on the environment. Certain impact assessments of some GE crops are also monitored even after deregulation.

Herbicide-tolerant (HT) Crops Lead to Super Weeds

Development of herbicide-tolerant weeds has occurred with both traditionally-bred and GE crops. This phenomenon reduces the effectiveness of certain weed control strategies and decreases weed management options.

Strategies have been developed to minimize the development of herbicide tolerant weeds, such as:

- Use of HT cultivars with resistance genes for herbicides with alternative modes of action that can be used in rotation.
- Use of restriction technologies to prevent gene passage to the next generation through the pollen, *i.e.* transgenes can be targeted to the cytoplasmic organelles, not in the pollen.
- Rotate the use of HT crops with different modes of action or with non HT crops.

A few points to consider in using HT crops are: Weeds can also escape herbicide treatment on the basis of application rate, weed age and size, spray volume adjuvants used, water quality and interactions with other herbicides that affect efficacy. Late germination of weeds can also escape herbicide application, thus a second pass of sprays can be done.

Effect of Using GE Crops in Pesticide Usse

Having crops tolerant to herbicides and pest attack increases pest management options and can also reduce the number and strength of pesticide applications. Growth of GE HT crops also allows topical application of herbicide to crops and weeds, which replaces spraying between crop rows and mechanical removal of weeds, both of which can damage crops and result in environmental damage. Reducing mechanical tillage lowers fuel consumption and helps conserve soils prone to erosion and compaction. HT crops can also lead to more flexible herbicide treatment regimes. The National Center for Food and Agricultural Policy published surveys on U.S. pesticide usage on GE crops. In 2004, HT canola, cotton, maize and soybean as well as Bt cotton and maize showed reductions in herbicide active ingredient (AI) of 25 to 30%.

In a 2006 publication, the USDA National Statistics Service found that from 1996 to 2002, AI use rates for HT cotton and corn, and Bt corn declined as adoption of Bt and HT cotton, corn, and soybeans increased and concurrent shifts occurred towards less environmentally persistent herbicides such as pendimethalin, trifluralin, and metolachlor. The Environmental Impact Quotient (IEQ) assessment which takes into account the pesticide AI and the environmental impact (EI) of GE crops resulted in significant reductions in the global EI of production agriculture; such that since 1996, the overall EI associated with pesticide use on HT soybean, corn, cotton, canola, and Bt cotton decreased by 15.3%. Cultivation of GE HT crops has also had other positive effects on the environment, *i.e.* increases in low-or no-till practices and use in combination with integrated pest management schemes, which were made possible because early season pesticide sprays could be eliminated, allowing beneficial insects to establish. Most reports indicate pesticide use and cost decrease following adoption of Bt varieties. In Argentina, numbers of herbicide applications increased with HT soybean but use shifted to more environmentally friendly herbicides. Reduction in pesticide use can also be achieved by using the best methods and tools available, including integrated pest management, biocontrol, organic production methods, and GE organisms to reduce EI while achieving adequate production levels.

Would Bt Crops Need Additional Insecticide Applications

Bt or Cry toxins are toxic to susceptible larvae when cleaved to generate their active form, which then binds to specific receptors in the midgut and creates holes that cause lepidopteran larvae to die. The first BT GE crops introduced into corn and cotton were targeted to control European corn borer, corn rootworm and cotton armyworm.

Some pests belong to groups insensitive to Bt have to be sprayed to prevent crop damage. With the commercial introduction of corn and cotton varieties with two stacked Bt genes, *i.e.* Cry1Ac and Cry2Ab in cotton, bollworms and secondary armyworm pests were controlled. New developments to target different insect pests are: corn with six insect resistant genes against lepidopteran (Cry1F, Cry1A.105, Cry2Ab2) and rootworm (Cry34Ab1 + Cry 35Ab1, modified by Cry3Bb1) pests; the use of a hybrid Cry protein with two binding domains to target lepidopteran and coleopteran pests of potato; use of plant defence proteins such as alpha amylase inhibitors from legumes; use of insecticidal compounds from nematodes, bacterial cholesterol oxidase, avidin, volatile communication compounds, and RNAi approaches targeted to specific insect proteins.

Even with GE approaches, other methods of insect control will be needed, *e.g.*, chemical pesticides, biocontrol, integrated pest management, or organic approaches, because insects are plentiful and ever changing.

Virus–Resistant Genetically Engineered Plants Lead to Novel Viruses

Development of GE crops with resistance to viral diseases has been conducted in squash and papaya using a viral coat protein gene. The USDA APHIS has already deregulated the GE squash allowing commercial production after the virus was shown not to infect wild squash varieties; the resistance gene gave no advantage to wild squash varieties, and the presence of the coat protein gene did not increase viral competitiveness. For GE papaya with the viral coat protein, concerns on viral recombination became a concern since from analyses of viruses, homologous and non homologous recombination could occur between viruses and between viral genomes and plant genes. Experimental results indicate however that most recombinant viruses are not fully virulent because the new gene combinations are not fully compatible, leaving new hybrids at a competitive disadvantage.

To compete effectively, recombinant viruses must have functional recombinatorial ability, capacity to establish systemic infection, and ability to compete with their progenitors during replication. These requirements place powerful negative selection pressure on newly evolved viruses. Reduced viral replication capacity could also negatively affect recombination frequency in transgenic plants. Large-scale field releases of plants engineered with viral genes are necessary to obtain realistic assessments of the types and recombination frequencies that might occur. Currently, no novel viruses have been reported resulting from GE plants in the field, but likely they would be detected only if their appearance had adverse effects.

At present, the only commercially propagated plants engineered with viral coat protein genes, GE squash, and papaya are grown on small acreages. To minimize the possibility for gene exchange among the viruses, strategies such as RNAi-mediated viral resistance is employed. There is no protein introduced, and the RNAi construct is used to silence a gene from bean golden mosaic virus in Phaseolus vulgaris leading to virusresistant plants.

Genetically Engineered Plants Move to Bacteria in the Field

Horizontal gene transfer is the process of transferring genes among nonsexually related organisms such as from plants to bacteria. Sequence analyses of genes and proteins show that some genes have transferred from plants to bacteria over a very long evolutionary time frame. This transfer can only be simulated in the laboratory using optimized conditions—situations which are difficult to replicate in natural settings. If, however it were to happen in the field, it would be at very low frequencies and the gene would need to provide a selective advantage to survive.

An experiment to determine the persistence of kanamycin resistant bacteria in the soil by increasing the levels of kanamycin in the soil concludes that natural soil conditions rarely would have the selective pressure necessary to keep nptII in the bacterium. Data from this and other studies indicate that homologous recombination and integration of plant genes into competent soil bacteria could occur, but at very low frequencies, and the environmental significance would depend on selective pressure for the trait.

When Pollen Moves From Genetically Engineered Crops to Wild Relatives or Non-genetically Engineered Varieties

Gene flow or the movement of pollen from one plant to another is made possible when the parental plants:

- Flowers at the same time;
- Close enoussgh to allow a vector (insect, wind, or animal) to transfer pollen to receptive females;
- Produce pollen that can result in embryos developing into viable seeds and germinating.

Successful pollination also depends on the longevity of pollen viability, pollen travel distance and the mode of pollination the plant has, whether self or cross-pollinated. Gene flow may present significant economic or environmental risks for either conventionally bred or GE crops on a case-by-case evaluation. Crop-to-wild relative gene flow could result if the plants grow in overlapping regions resulting in new combinations of genes that can improve, harm, or have no effect on the fitness of recipient plants.

Genes can also flow from wild relatives to cultivated crops, introducing new traits into the next generation seeds, but only affect the crop if it is replanted. Planting of GE varieties in areas of genetic diversity of plants needs additional precautions to reduce possible impacts of introgression of GE traits and the potential significant environmental consequences.

To minimize this occurrence, planting of GE crops near wild species should be avoided or other technologies could be used to prevent gene(s) from moving to wild varieties. Gene flow could also occur when compatible plants are present within the vicinity. GE varieties like conventional plants can also persist in the environment. Organic farmers should be aware of these occurrences to be able to adopt the necessary precautions of spatial and temporal isolation.

Organic, Conventional and Genetically Engineered Cropping Systems Coexist

Farmers are used to planting different varieties and planting strategies in order to develop farm products that meet the requirements of the consumers. They are used to planting white and yellow maize, hot and sweet peppers, high and zero erucic acid rapeseed, and still achieve purity standards dictated by certified seed specification. Coexistence strategies must be devised to allow neighbour farmers to farm in an economically viable manner. This can involve alerting each other to their plans and modifying them to accommodate each others' needs. When GE crops are grown next to organic farming operations, certain practices that minimize synthetic pesticide drift can also limit GE gene flow, such as spatial separation of fields, staggered planting dates, and planting varieties with different maturity dates and those that are not sexually compatible.

Other crops-specific methods have been devised to aid coexistence strategies. Gene flow is not only the means for GE to commingle with conventional or organic crops; crops must also be segregated during harvest, shipping and processing. Methods limiting such commingling have in some cases been implemented. With the use of various production methods comes the mixing of permissible inputs and methods, whether with their own farms with products from neighbouring farms, or during harvest and processing. The commingling or adventitious presence (AP) is the unintended occurrence of materials other than specific crops and can include weed seeds, seeds from other crops, dirt, insects, and other foreign materials such as stones or plastics. Different countries have set rules on the degree of AP. In the U.S., for seed crops, rules for AP are specified by the Association of Official Seed Certifying Agencies (AOSCA), where a level of 0.5% seed of other varieties and 2% AP of inert materials is permitted in "pure seed" of hybrid corn.

Venetically Engineered Crops or Organic Farming Lead to More Sustainable Agricultural Production Systems

Sustainable agricultural systems should meet the basic needs of the population while preserving the resources for future generations. The United Nation's Millennium Development Goals to "Ensure environmental sustainability by integrating principles of sustainable development into a country's policies, and programs to reverse the loss of environmental resources."

This need has been widely accepted and the manner to fulfill this may vary. Conventional farming has led to impressive gains of between 70 and 90% of increases in food production in the last few decades. Unfortunately, these were accompanying environmental impacts as well as sizeable consumption of fossil fuels, unsustainable rates of water use and topsoil loss, and contributions to environmental degradation, air pollution, soil erosion, reduced biodiversity, pest resistance, pollution of lakes and streams, and overuse of surface and ground water. Achieving agricultural sustainability can be addressed through numerous agricultural practices such as: integrated pest management (IPM), biological control, organic methods, and use of GE plants, coupled with selected conventional agricultural methods, can play important roles in future sustainable agricultural practices. Biological control can be a part of an IPM strategy and neither biological control nor IPM specifically excludes the use of GE organisms. Organic production relies on practices, such as cultural and biological pest management, that can include IPM and biological control but excludes the use of synthetic chemicals and GE organisms.

The use of GE organisms can also contribute to sustainable practices by augmenting and replacing certain conventional practices. For example, plants can be created that increase water use, and fertilizer efficiencies, that remediate soil contaminants, increase no-till or low-till practices to help reduce greenhouse gases and produce higher yields without increasing land usage, particularly in developing countries. To achieve true sustainability agriculture must use the best of all practices.

9

Economic Perspectives of Agricultural Biotechnology

Science-based regulation of agricultural biotechnology contributes to the free trade of safe biotech applications and to the appropriate use of this technology to promote development, writes Alan Larson, under secretary of state for economic, business and agricultural affairs. Larson adds that biotechnology—one of the most promising new technologies of our times—is too important for the future prosperity of the world to ignore.

Biotechnology is one of the most promising new technologies of our times. The expanding use and trade of agricultural biotechnology-derived products is enhancing prosperity and well-being both in developed and developing countries. Unfortunately, while the United States and many other nations around the world are expanding the development and use of safe biotechnologyderived products, some countries have imposed unjustified restrictions on them. Such restrictions threaten the international trading system and are preventing developing countries from exploring the enormous potential of biotechnology to improve the lives of their people.

BIOTECHNOLOGY AND DEVELOPMENT

In 2000, the world's population was about 6 billion. It is expected to increase to 9 billion by 2050. As a result, there will be more people to feed on an increasingly crowded planet. Food production will have to increase, and it must increase in an environmentally sustainable way. Since 1980, 50 per cent of the increased agricultural productivity in the developing world came through improved seed technology. Better seeds can come from improving traditional methods, developing conventional hybrids, and through biotechnology. Biotechnology, while not a panacea, can make an important contribution. Agricultural biotechnology achieves enhanced crop productivity in a more environmentally sustainable way. In the United States, the growing use of agricultural biotechnology is resulting in reduced use of pesticides and increased adoption of environmentally friendly farming practices such as "no-till" farming, which reduces soil erosion and fertilizer run-off. Enhanced productivity means that more food can be raised on the same amount of land.

As population pressure grows in the coming years, the ability to grow enough food for the world's burgeoning population without encroaching on vital habitats such as tropical rainforests will be of enormous benefit to the environment. The United States is not the only country that is reaping the benefits of biotechnology. New crops derived from biotechnology are being used in developing countries such as Argentina, South Africa, China, the Philippines and India.

The attraction of biotechnology in these countries lies in the direct benefits these varieties bring to the developing country farmer. In China, for example, where small farmers grow biotechnology-derived insectresistant cotton varieties in great numbers, these varieties require fewer pesticides, which not only reduce costs, but also significantly reduce exposure to dangerous chemicals.

As a result, farmers are healthier and have expanding incomes that let them buy better food for their families or send a child to school rather than have that child work in the fields. Such results, spread over the population of an entire country where farmers are by far the largest percentage of the population, provide the opportunity for development and improved prosperity. The challenge is to make tried and tested biotechnology varieties available to more developing countries and to help develop new varieties specifically adapted for their conditions. This is why the United States supports the development of biotechnology-derived staple food crops that will fight disease such as insect-resistant cowpeas, disease-resistant bananas, cassava and sweet potatoes. Biotechnology may also offer a quicker route for undernourished populations to get access to a better diet. For example, a Vitamin A enriched rice variety known as "golden rice" is under development to help fight blindness caused by malnutrition.

The potential benefits of this new technology should not be thrown away or delayed unnecessarily. Last year a few African nations balked at receiving badly needed food aid—food most Americans eat every day—because of unscrupulous and unscientific fear mongering. This must stop. Rather, the international community should reach out to developing countries—as the United States is doing—to explain how safe biotechnology-derived products can be regulated, used domestically, and traded abroad to the benefit of all.

BIOTECHNOLOGY AND TRADE

Despite the benefits of biotechnology for both the developed and developing world, biotechnology-derived crops are at the center of a number of contentious trade disputes. This is the case even though more than 3,200 esteemed scientists around the world—including 20 Nobel

Laureates—have concluded that the biotechnology-derived products currently on the market do not pose greater risks to human health than their conventional counterparts.

The only way to maintain a free and fair trading system is for products traded in that system to be regulated in a logical, objective and science-based manner. When such a system is in place, we can have confidence in the safety of the products we trade. How biotechnology-derived crops are treated in the international system will have consequences not just for biotechnology, but also for all new technologies. It is important that we get this right. The rules governing the trade of biotechnology-derived products, and indeed all products, must be based on scientific risk assessment and risk management. The World Trade Organization (WTO) Agreement on Sanitary and Phytosanitary Measures (SPS Agreement) requires that measures regulating imports be based on "sufficient scientific evidence" and that countries operate regulatory approval procedures "without delay." When science is the basis of decision-making, countries find it easier to agree on rules. For example, the Codex Alimentarius Commission recently approved science based guidelines for biotechnology food safety assessments relating to human health.

These guidelines were approved unanimously by the Commission, which is composed of 169 members, including the U.S., EU (European Union) member countries, and the vast majority of developing nations. Three international standard setting bodies, including Codex, are specifically recognized by the WTO SPS Agreement. The Codex Alimentarius Commission develops food safety standards.

The International Plant Protection Convention (IPPC) focuses on preventing the spread and introduction of pests in plants and plant products. The Office of International Epizootics (OIE) performs a similar function for animal health. All three organizations base their work on scientific analysis. It is essential for the integrity of the international trading system that the WTO continue to refer to the work of these bodies in assessing biotechnology products and that these organizations continue to perform science-based work.

The U.S. supports workable, transparent and science based regulations for agricultural biotechnology applications. In fact, the U.S. government provides technical assistance to countries to help them develop their own capacity to regulate this technology and put it to use for the benefit of their citizens. When countries adopt a science-based approach to biotechnology, fair rules for the regulation and trade of biotech products can be established. The U.S. is committed to pursuing such a science-based approach to biotechnology with its trading partners and is convinced that this approach is the best way to ensure a fair and safe trading system for agricultural biotechnology products. Agricultural biotechnology can help both the developing and developed world enhance productivity while preserving the environment. Science-based regulation of agricultural biotechnology applications contributes to the free trade of safe biotech applications and to the appropriate use of this technology to promote development. Scientists around the world, including those in the European Union, agree that there is no evidence that approved biotechnology-derived foods pose new or greater dangers to the environment or to human health than their conventional counterparts. Indeed, any alleged downsides to agricultural biotechnology lie in the realm of the theoretical and potential. The upsides have already been demonstrated. Biotechnology is too important for the future prosperity of the world to ignore.

AGRICULTURAL BIOTECHNOLOGY AND THE DEVELOPING WORLD

Biotechnology has the potential to play a large role in more rapidly advancing agricultural productivity in developing countries while protecting the environment for future generations, writes J.B. Penn, under secretary for farm and foreign agricultural services at the U.S. Department of Agriculture. Penn says biotechnology is simply another crop improvement tool in the long history of cultivation. Agricultural biotechnology has been changing the face of agriculture since its commercial introduction in 1996 and the widespread adoption of bioengineered crops by farmers in the United States and other countries.

However, this technology is not without controversy and is causing political reverberations around the world. While it holds enormous promise for significantly increasing food production and relieving already strained land and water resources, it has become an emotional issue among some consumers and environmental groups. As the science continues to be developed, it clearly will present both opportunities and challenges to participants throughout the food chain.

BACKGROUND ON CONVENTIONAL PLANT BREEDING

Almost all plants can be considered "genetically modified." Genetic modification occurs when plants within a species simply produce offspring. The offspring is not exactly like either of the parents; it is a genetic combination of both. For centuries, plants have been cultivated and cross-bred by man to produce offspring with specific, desired traits. For example, maize as we know it today barely resembles its ancestor, teosinte, or Zea mexicana, a tall grass that produces finger-length "ears" containing a single row of a few grains. Maize produced today has been cultivated for many years to serve as a food crop, with far different traits than those of its predecessors. When varieties are cross-bred to produce a hybrid plant, millions of genes are combined in the process. Scientists must select and continually cross-breed the plants, often over a period of several years, to obtain plants with the largest number of desired traits and the least number of undesirable traits.

Modern biotechnology

Modern biotechnology is a tool that allows scientists to select a single gene for a desired trait, incorporate it into plant cells, and grow plants with the desired trait. In many ways it is simply a "high-tech" version of traditional plant breeding. This more efficient process prevents millions of genes from being crossed and possibly producing undesirable traits. Biotechnology is also different because it allows scientists to incorporate genes from other species—something that cannot be done via conventional plant breeding.

This makes biotechnology a very powerful and useful tool for plant breeders. Some people fear this tool because it is perceived as "unnatural." However, most people forget that the food crops we have today would not exist without man's intervention, whether through plant breeding, fertilizer application, delivery of irrigation water or use of modern tractors and equipment.

Without cultivation by man over the years, we would still have teosinte instead of conventional maize. The same is true for wheat, tomatoes, potatoes, watermelon and any product on today's supermarket shelf. Thus, biotechnology is simply a modern, additional tool in the long history of plant cultivation and agriculture.

AGRICULTURAL BIOTECHNOLOGY TODAY

While the focus of the first "generation" of biotech crops has been on the considerable economic benefits to farmers, more and more evidence is accumulating that significant food safety and environmental benefits are beginning to accrue. Farmers have indicated their acceptance of biotech varieties by the unprecedented pace in which they have been adopted. According to the U. S. Department of Agriculture (USDA), in the United States approximately 80 per cent of soybeans, 38 per cent of maize and 70 per cent of cotton were planted to biotech varieties in 2003. The United States is not alone in experiencing this evolution in agriculture. Adoption rates in other countries, such as Argentina, Canada and China, where biotech varieties are approved, have been similarly rapid.

According to the National Center for Food and Agricultural Policy in Washington, D.C., U.S. farmers have realised the following benefits through the use of biotech varieties:

- *Roundup Ready Soybeans*: 28.7 million lbs. (13,018.3 metric tons)/ year decrease in herbicide use; $1.1 billion/year savings in production costs.
- *Bt Cotton*: 1.9 million lbs. (861.8 metric tons)/year decrease in insecticide use; 185 million lbs. (83,916 metric tons)/year increase in cotton production.
- *Bt Maize Varieties*: Over 16 million lbs. (7,257.6 metric tons)/year decrease in insecticide use; 3.5 billion lbs. (1,587,600 metric tons)/ year increase in production volume.
- *Papaya*: Virus-resistant biotech papaya saved the Hawaiian papaya industry $17 million/year in 1998 from the devastating effects of ringspot virus. These results illustrate enormous decreases in pesticide use, with corresponding environmental enhancement, along with equally dramatic increases in production and savings in production costs. While biotech results vary by farm, the economic benefits obviously have been significant. These benefits are realised not only by farmers, but also by the environment and to consumers in general.
- The reduced reliance of biotech varieties on chemical inputs means less water pollution.
- Reduced chemical usage results in safer water supplies and higher quality drinking water as well as a better environment for wildlife.
- Higher yielding biotech crops can help ease the strain on land resources, reducing the need for expansion onto more fragile areas and thus allowing for greater conservation of natural habitats.
- Energy usage on biotech crops is lower because there are fewer passes through fields in applying chemicals. Less fuel use means less carbon entering the atmosphere as carbon dioxide (CO_2).
- Herbicide-resistant crops encourage the adoption of conservation tillage, especially no-till, which reduces erosion of topsoil.

Current research ON Modern biotechnology

Current research will lead to food crops that are resistant to environmental pressures such as drought, temperature extremes and salty soil. Scientists around the world are also investigating the "second generation" of biotech products - those with direct consumer benefits such as enhanced nutrition levels.

Many of us have heard of "golden rice," which contains higher levels of beta carotene–an important component in vitamin A production. Scientists in India are working to develop a biotech potato variety with higher levels of protein. Edible vaccines could also be produced by plants to provide low-cost, low-maintenance medicines. These are just a few of the numerous examples of cutting edge research that will further the changes we have already witnessed in the global food chain. The possibilities are enormous.

IMPLICATIONS FOR THE DEVELOPING WORLD

Global population projections suggest an additional 725 million mouths to feed in just 10 years. By 2020, this will grow to 1.2 billion more people to feed—equivalent to the populations of all Africa and South America combined. This expansion comes despite the fact that today some 800 million people - nearly one in seven—face chronic hunger. This is especially devastating to the world's children, where one in three is undernourished, and a child dies every five seconds due to hunger. Biotechnology alone will not feed tomorrow's world. However, this far-reaching agricultural technology, in combination with political and economic reforms, can increase crop productivity by increasing yields and improving the nutritional content of crops in developing countries. It will also help provide lower-cost food to low-income consumers. Bringing such benefits to developing countries would have far-reaching results, indeed. An annual increase of 3 to 4 per cent in African crop and livestock yields would almost triple per capita incomes while reducing the number of malnourished children 40 per cent.

Increased agricultural productivity will drive economic growth and expand opportunities to trade, bringing more and better jobs, better health care, and better education. Consumers in developing countries spend a high proportion of their disposable income on food, which could be reduced with a more efficient food system, thereby leaving more of their income for other products to enhance their quality of life.

The most critical areas in the world for bringing economic prosperity and stability are the developing countries. Agricultural productivity in these countries must advance more rapidly to meet growing food demand and raise incomes while protecting the environment for future generations. Biotechnology has the potential to play a large role in this achievement.

UNDERSTANDING BIOTECHNOLOGY IN AGRICULTURE

Bioengineering provides distinct advantages over traditional breeding technologies because the risk of introducing detrimental traits is likely to be reduced, says Deputy U.S. Food and Drug Administration Commissioner Lester Crawford. Crawford, a doctor of veterinary medicine by training, argues that there are no scientific reasons that a product should include a label indicating that it, or its ingredients, was produced using bioengineering.

He also outlines draft guidelines to strengthen controls that would prevent biotech products in field trials from inadvertently getting into food or feed. Based on two decades of experience with bioengineered foods and overwhelming scientific data that these foods are safe to eat, we believe that biotechnology can offer a safe and important tool for both exporting and food deficit countries. This paper describes some of the basic science behind biotechnology, the U.S. regulatory structure for ensuring safe foods and U.S. policy on the issue of labeling.

CROSS-BREEDING, HYBRIDIZATION AND BIOENGINEERING

Scientists have been improving plants by changing their genetic makeup since the late 1800s. Typically, this has been accomplished through cross-breeding and hybridization, in which two related plants are cross fertilized and the resulting offspring have characteristics of both parent plants. In the breeding process, however, many undesirable traits often can appear in addition to the desirable ones. Some of those undesirable traits can be eliminated through additional breeding, which is time consuming.

Breeders can then further select and reproduce the offspring that have the desired traits. Many of the foods that are already common in our diet are obtained from plant varieties that were developed using conventional genetic techniques of breeding and selection. Hybrid maize, nectarines, which are genetically altered peaches, and tangelos, which are a genetic hybrid of a tangerine and grapefruit, are all examples of such breeding and selection. Today, by inserting one or more genes into a plant, scientists are able to produce a plant with new, advantageous characteristics. The new gene splicing techniques are being used to achieve many of the same goals and improvements that plant breeders historically have sought through conventional methods. They give scientists the ability to isolate genes and introduce new traits into foods without simultaneously introducing undesirable traits. This is an important improvement over traditional breeding. Because of the increased precision offered by the bioengineered methods, the risk of introducing detrimental traits is actually likely to be reduced.

FOOD SAFETY CONCERNS

The U.S. Food and Drug Administration (FDA) have found no evidence to indicate that either ordinary plant deoxyribonucleic acid (DNA) or the DNA inserted into plants using bioengineering presents food safety problems. Nor are the small amounts of the newly expressed proteins likely to change dramatically the safety profile of the plant. If safety concerns should arise, however, they would most likely fall into one of three broad categories: allergens, toxins or anti-nutrients. FDA has extensive experience in evaluating the safety of such substances in food.

It is important to note that the kinds of food safety testing typically conducted by developers of a bioengineered food crop to ensure their foods meet all applicable requirements of the Food, Drug and Cosmetics Act (FD&C Act) address these potential concerns. In the event that something unexpected does occur, this testing provides a way to detect such changes at the developmental stage and defer marketing until any concern is resolved.

As aforementioned, some of the food safety concerns that could arise include:

- *Allergens:* Foods normally contain many thousands of different proteins. While the majority of proteins do not cause allergic reactions, virtually all known human allergens are proteins. Since genetic engineering can introduce a new protein into a food plant, it is possible that this technique could introduce a previously unknown allergen into the food supply or could introduce a known allergen into a "new" food.
- *Toxins:* It is possible that a new protein, as introduced into a crop as a result of the genetic modification, could cause toxicity.
- *Anti-nutrients:* It is possible that the introduction of anti-nutrients, such as molecules like phytic acid, could reduce essential dietary minerals such as phosphorus. The use of genetic engineering techniques could also result in unintended alterations in the amounts of substances normally found in a food, such as a reduction of Vitamin C or an increase in the concentration of a naturally occurring toxicant in the plant food.

LEGAL AND REGULATORY ISSUES

One important component in ensuring food safety is the U.S. regulatory structure. The FDA regulates bioengineered plant food in conjunction with the United States Department of Agriculture (USDA) and the Environmental Protection Agency (EPA). FDA has authority under the FD&C Act to ensure the safety of all domestic and imported foods for man or animals in the United States market. The exceptions to this are meat, poultry and certain egg products, which are regulated by USDA.

The safety of animal drug residues in meat and poultry, however, is regulated by FDA. Pesticides, including those bioengineered into a food crop, are regulated primarily by EPA. USDA's Animal and Plant Health

Inspection Service (APHIS) oversees the agricultural and environmental safety of planting and field testing bioengineered plants. Bioengineered foods and food ingredients must adhere to the same standards of safety under the FD&C Act that apply to their conventionally bred counterparts.

This means that these products must be as safe as the traditional foods in the market. FDA has the power to remove a food from the market or sanction those marketing the food if the food poses a risk to public health. It is important to note that the FD&C Act places a legal duty on developers to ensure that the foods they market to consumers are safe and comply with all legal requirements.

FOOD ADDITIVES

A substance that is intentionally added to food is a food additive, unless the substance is generally recognized as safe (GRAS) or is otherwise exempt, such as a pesticide whose safety is overseen by EPA. The FD&C Act requires premarket approval of any food additive regardless of the technique used to add it to food. Thus, substances introduced into food are either new food additives that require premarket approval by FDA, or GRAS and are therefore exempt from the requirement for premarket review. Generally, foods such as fruits, vegetables and grains are not subject to premarket approval because they have been safely consumed over many years. Other than the food additive system, there are no premarket approval requirements for foods generally. Under FDA policy, a substance that would be a food additive if it were added during traditional food manufacturing is also treated as a food additive if it is introduced into food through bioengineering of a food crop. Our authority permits us to require premarket approval of any food additive and, thus, to require premarket approval of any substance intentionally introduced via bioengineering that is not generally recognized as safe.

Examples of substances intentionally introduced into food that would be reviewed as food additives include those that have unusual chemical functions, have unknown toxicity, or would be new major dietary components of the food. For example, a novel sweetener bioengineered into food would likely require premarket approval. In our experience with bioengineered food to date, however, we have reviewed only one substance under the food additive provisions, an enzyme produced by an antibiotic resistance gene, and we granted it approval as a food additive. In general, substances intentionally added to or modified in food via biotechnology to date have been proteins and fats that are, with respect to safety, similar to other proteins and fats that are commonly and safely consumed in the diet and, thus, are presumptively GRAS. Therefore, they have not needed to go through the food additive approval process.

PRE-MARKET CONSULTATIONS

We are not aware of any bioengineered plant food that is subject to FDA's jurisdiction and is on the market that has not been evaluated by FDA through the current consultation process. Typically, the consultation begins early in the product development stage, before the product is ready for market. Company scientists and other officials meet with FDA scientists to describe the product they are developing.

The agency then advises the company on what tests would be appropriate for the company to assess the safety of the new food. After the studies are completed, the data and information on the safety and nutritional assessment are provided to FDA for review. FDA evaluates the information for all of the known hazards and also for potential unintended effects on plant composition and nutritional properties since plants may undergo changes other than those intended by the breeders. For example, FDA scientists are looking to assure that the newly expressed compounds are safe for food consumption and that there are no allergens new to the food, no increased levels of natural toxicants, and no reduction of important nutrients.

They are also looking to see whether the food has been changed in any substantive way such that the food would need to be specially labeled to reveal the nature of the change to consumers. If a plant developer used a gene from a source whose food is commonly allergenic, FDA would presume that the modified food might be allergenic.

The developer, however, is allowed the opportunity to demonstrate that such food would not cause allergic reactions in persons allergic to food from the source. Our experience has been that no bioengineered product has gone on the market until FDA's questions about the safety of the product have been answered.

LABELING

One of the most important issues confronting the biotechnology industry is that of labeling. Under the FD&C Act, a food is misbranded if its labeling is false or misleading in any particular way. FDA does not require labeling to indicate whether or not a food or food ingredient is a bioengineered product, just as it does not require labeling to indicate which conventional breeding technique was used in developing a food plant. However, if genetic modifications materially change the composition of a food product, these changes must be reflected in the food's labeling.

This would include its nutritional content (for example, more oleic acid or greater amino acid or lysine content) or requirements for storage, preparation or cooking, which might impact the food's safety characteristics or nutritional qualities. For example, one soybean variety was modified to alter the levels of oleic acid in the beans. Because the oil from this soybean is significantly different from conventional soybean oil, we advised the company to adopt a new name for that oil, a name that reflects the intended change.

If a bioengineered food were to contain an allergen not previously found in that food and if FDA determined that labeling would be sufficient to enable the food to be safely marketed, FDA would require that the food be labeled to indicate the presence of the allergen. FDA has received comments suggesting that foods developed through modern biotechnology should bear a label informing consumers that the food was produced using bioengineering.

We have given careful consideration to these comments. However, we do not have data or other information to form a basis for concluding that the fact that a food or its ingredients were produced using bioengineering constitutes information that must be disclosed as part of a bioengineered product's labeling. Hence, we believe that we have neither a scientific nor a legal basis to require such labeling. We have developed, however, draft guidance for those who wish voluntarily to label either the presence or absence of bioengineered food in food products.

STRENGTHENING CONTROLS OVER FIELD TRIALS

In August 2002, President Bush's Office of Science and Technology Policy (OSTP) proposed strengthening controls over field trials to address the potential of material from field trials inadvertently getting into food or feed. FDA's task is to publish draft guidance for comment on procedures to address the possible intermittent, low-level presence in food and feed of new non-pesticidal proteins from biotechnology-derived crops that are under development for food or feed use but have not gone through FDA's premarket consultation process.

Under this guidance, FDA would encourage sponsors, domestic and foreign, to submit protein safety information when field testing showed that there could be concerns that new non-pesticidal proteins produced in the field-tested plants might be found in food or feed. FDA's focus would be on proteins new to such plants because FDA believes that at the low levels expected from such material, any food or feed safety concerns would be limited to the potential that a new protein could cause an allergic reaction in some people or could be a toxin.

PHARMACEUTICAL CROPS

FDA has the authority and responsibility for regulating pharmaceuticals, whether they are manufactured in a traditional manufacturing plant or manufactured in crops in the field. For crops in the field, however, there are additional issues to be addressed, including issues involving the parts of the plant that do not contain the pharmaceutical and the residual crop left over after a pharmaceutical is extracted. In September 2002, FDA and USDA published Draft Guidance for Industry on the use of bioengineered plants or plant materials to produce biological products, including medical devices, new animal drugs, and veterinary biologics. This draft guidance outlines the important scientific questions and information that should be addressed to FDA by those who are using bioengineered plants to produce medical or veterinary products. We are currently reviewing public comments on this guidance.

After 10 years of experience in this country, there is every reason to conclude that bioengineered foods are as safe as food produced through traditional breeding techniques. Both the U.S. General Accounting Office (GAO) and the National Academy of Sciences (NAS) have issued reports agreeing with this assessment. We are confident that the foods developed using bioengineering that we have evaluated are as safe as their counterparts, and we will continue to follow the development of this technology to ensure that any new safety questions are also resolved prior to marketing.

A GREEN FAMINE IN AFRICA

Facing famine must consider the severe, immediate consequences of rejecting food aid that may contain biotechnology, writes Tony Hall, U.S. Ambassador to the U.N. Agencies for Food and Agriculture. Southern African countries that faced severe food shortages in late-2002 and rejected U.S. food aid, risked the lives of millions of their people. The rejected food, he writes, is the same food people in the United States eat and has undergone rigorous food safety and environmental impact testing. Last year and the first few months of 2003, Southern Africa was on the verge of a catastrophe.

It was on the brink of famine and is not out of the woods yet. The United States Government did everything we could to stop it and, for the most part, we were successful. The causes were, and remain, varied: drought, a rampant HIV/AIDS epidemic that orphans millions and failed governments prepared to play the politics of hunger. Some governments even blocked the delivery of emergency food relief needed to head off starvation. Their excuse was derived from the ongoing debate over biotechnology, spurred in part by certain European bias against biotechnology. Last October, I went to visit Zimbabwe and Malawi, two of the six nations affected by the crisis. As the newly arrived U.S. Ambassador to the United Nations Agencies for Food and Agriculture, I had to see this crisis first hand. After almost 24 years of fighting hunger as a U.S. Congressman, however, I had a good idea of what famine looked like. I visited hospitals, feeding centers and schools. I saw many malnourished people—mostly children—and when I asked these children "when is the last time you ate?" most replied that it had been two days, and some said five or six days. Hospitals were overflowing with children they struggled to keep alive.

This is another result of the HIV/AIDS epidemic that has created almost one million orphans in Zimbabwe alone, and perhaps 800,000 in Malawi, with no means of support or sustenance. U.S. and international experts agreed that the worsening food crisis in southern Africa placed as many as 14.5 million people at risk. These people did not have enough food then and most do not have enough today. Hunger continues to haunt many of their days.

Even though we have done much to assist, they are in different stages of starvation. The situation in Zimbabwe is still headed for major disaster. Zambia could have been even worse. In 2001, the U. S. Famine Early Warning System (FEWSNET) identified the onset of drought and food shortages. By February 2002, the United States was moving emergency relief into the region with the World Food Programme (WFP). In southern Africa, more than 350 thousand metric tons of U.S. food aid had been delivered by November and another 150 thousand metric tons were delivered in the following three months.

This still represented only half the food the region needed. But food that should have gotten into Zimbabwe and Zambia with ease was stuck outside these countries, while debate raged inside over the human health and environmental risks posed by the maize millions of Americans eat daily. Moreover, the Zambian government decided to reject the maize the U.S. had donated. More than 15,000 tons of U.S. maize had to be removed from the country by WFP at a cost of almost $1 million. There were riots when some hungry Zambian citizens learned of their government's plan and some of the food eventually made it back into the country through the black market. It doesn't take a lot to calculate the impact of these debates, carried out by well-fed experts. As the region headed for famine, vulnerable people perished. While the U.S. respects the rights of countries to make their own decisions about biotechnology, we have no other option but to provide the food we consume ourselves. And other donors simply could not have increased their donations to fill the gap had more U.S. food aid been rejected.

The United States provides between one-half and two-thirds of the food aid needed to meet emergencies around the world. All of this food comes from our own stocks and markets. It is the same food we eat. It is the same food we feed our children. Maize is the staple food of southern Africa and U.S. maize is about one-third biotech. All of the food donated by the United States has passed our rigorous food safety and environmental impact testing.

In fact, it is eaten daily and has been for years by millions of Americans, Canadians and South Africans, and millions of other people all over the world. We have the most rigorous food safety testing system in the world. For this reason, U.S. biotech and non-biotech foods are mixed together. We do not, and see no need to separate them. At the request of Secretary General Kofi Annan, the World Food Programme, the World Health Organization (WHO) and the Food and Agriculture Organization (FAO) issued a joint policy on biotechnology in the summer of 2002. It stated that, based on all scientific evidence, genetically modified (GM)/biotech foods now marketed present no known risk to human health.

The European Commission also issued a public statement in August 2002, which agreed that there was no evidence that genetically modified maize varieties are harmful. Even strong biotech opponents such as Greenpeace belatedly recommended that African countries accept GM maize as an alternative to starvation. But years of anti-biotech lobbying, demands for a "precautionary principle" that no amount of science can satisfy, and a mistrustful climate provide a ready excuse. This climate is fostered in part by some nongovernmental organizations (NGOs) that seek to capitalize on repeated scares over food safety regulations in Europe that have nothing to do with biotech. When I was in Zimbabwe and Malawi, nobody asked me about the safety of biotech food. Nobody. Starving people, of course, simply want to be fed. But civil servants in the governments of Zimbabwe and Malawi did not ask, nor NGO relief workers, nor anyone else.

It is vitally important that the countries and the international community carefully consider new and emerging issues such as biotechnology. But it is also important that we realise that our actions, or our inactions, have consequences. People can die, they did die and they will die. The United States remains ready to help. Leaders in affected countries are, of course, free to choose whether to accept that help. But as Gro Brundtland, former head of the World Health Organization stressed, they must consider the severe, immediate consequences of rejecting food aid that is made available for millions of people so desperately in need. Time could run out.

THE CARTAGENA PROTOCOL ON BIOSAFETY

More than 130 countries adopted the Biosafety Protocol on January 29, 2000, in Montreal, Canada. It is called the Cartagena Protocol on Biosafety to honour Cartagena, Colombia, which hosted the extraordinary Conference of the Parties to the Convention on Biological Diversity (CBD) in 1999. The objective of this first Protocol to the CBD is to contribute to the safe transfer, handling and use of living modified organisms (LMOs)—such as genetically engineered plants, animals and microbes—that cross international borders. The Biosafety Protocol is also intended to avoid adverse effects on the conservation and sustainable use of biodiversity without unnecessarily disrupting world food trade. The Protocol will enter into force on September 11, 2003. Although the United States is not a Party to the CBD and therefore cannot become a Party to the Biosafety Protocol, the U.S. participated in the negotiation of the text and the subsequent preparations for entry into force under the Intergovernmental Committee on the Cartagena Protocol. We will participate as an observer at the first Meeting of the Parties (MOP1), scheduled for February 2004 in Kuala Lumpur, Malaysia.

The Protocol provides countries the opportunity to obtain information before new biotech organisms are imported. It acknowledges each country's right to regulate bio-engineered organisms, subject to existing international obligations. It also creates a framework to help improve the capacity of developing countries to protect biodiversity.

Biosafety Clearing-House

The Protocol establishes an Internet-based "Biosafety Clearing-House" to help countries exchange scientific, technical, environmental and legal information about living modified organisms (LMOs). It creates an advance informed agreement (AIA) procedure that in effect requires exporters to seek consent from an importing country before the first shipment of an LMO meant to be introduced into the environment, such as seeds for planting, fish for release or microorganisms for bioremediation.

It requires shipments of LMO commodities, such as maize or soybeans that are intended for direct use as food, feed or for processing, to be accompanied by documentation stating that such shipments "may contain" living modified organisms and are "not intended for intentional introduction into the environment." The Protocol establishes a process for considering more detailed identification and documentation of LMO commodities in international trade.

It also sets out information to be included on documentation accompanying LMOs destined for contained use, including any handling requirements and contact points for further information and for the consignee. The Protocol includes a "savings clause," which states that the agreement shall not be interpreted as implying a change in the rights and obligations of a Party under any existing international agreement, including, for example, World Trade Organization (WTO) agreements. The Protocol calls on Parties to cooperate with developing countries in building their capacity for managing modern biotechnology. The Protocol does not address food safety issues. Experts in other international fora, such as Codex Alimentarius, address food safety. It does not pertain to non-living products derived from genetically engineered plants or animals, such as milled maize or other processed food products. It does not require segregation of commodities that may contain living modified organisms. It does not subject commodities to the Protocol's AIA procedure, which would significantly disrupt trade and jeopardize food access, without commensurate benefit to the environment.

The Protocol does not require consumer product labeling. The mandate of the Protocol is to address risks to biodiversity that may be presented by living modified organisms. Issues related to consumer preference were not part of the negotiation. The Protocol's requirement for documentation identifying commodity shipments as "may contain living modified organisms" and "not intended for intentional introduction into the environment" can be accomplished through shipping documentation.

KEY PROVISIONS OF THE BIOSAFETY PROTOCOL

Advance Informed Agreement (Aia) Procedure

The Protocol's AIA procedure, in effect, requires an exporter to seek consent from an importing country prior to the first shipment of a living modified organism (LMO) intended for introduction into the environment, *e.g.*, seeds for planting, fish for release and microorganisms for bioremediation. The AIA procedure does not apply to LMO commodities intended for food, feed or processing, *e.g.*, maize, soy or cottonseed, to LMOs in transit, or to LMOs destined for contained use, *e.g.*, organisms intended only for scientific research within a laboratory. Importers are to make decisions on the import of LMOs intended for introduction into the environment based on a scientific risk assessment and within 270 days of notification of intent to export.

Commodity Requirements/ Biosafety Clearing-house

The agreement requires governments to provide the Biosafety Clearing-House with information concerning any final decisions on the domestic use of an LMO commodity within 15 days of making a decision. The agreement sets forth different shipping documentation requirements for different types of LMOs. These requirements will be in effect after the Protocol comes into force. Documentation accompanying shipments of LMOs intended for introduction into the environment, *e.g.*, seeds for planting, must identify the shipment as containing LMOs along with the identity and relevant traits and/or characteristics of the LMO, any requirements for safe handling, storage, transport and use, the contact point for further information, a declaration that the movement is in conformity with the Protocol and, as appropriate, the name and address of the importer and exporter.

Documentation accompanying shipments of LMO commodities intended for direct use as food or feed, or for processing, must indicate that the shipment "may contain" LMOs, that the shipment is not intended for intentional introduction into the environment, and specify a contact point for further information.

The Protocol provides for a decision by the Parties on the need for detailed requirements for this purpose, including specification of the identity and any unique identification of the LMOs, no later than two years after the entry into force of the Protocol. Documentation accompanying LMOs destined for contained use, *e.g.*, for scientific or commercial research within contained facilities, must identify the shipment as containing LMOs and must specify any requirements for safe handling, storage, transport and use, the contact point for further information, including the name and address of the individual and institution to whom the LMOs are consigned.

RIGHTS AND OBLIGATIONS UNAFFECTED

As evidenced by both the substantive content of the Protocol and its preambular "savings clause," Parties must implement rights and obligations under the Protocol consistent with their existing international rights and obligations, including with respect to non-Parties to the Protocol.

Precaution

Precaution is reflected in the Protocol's preamble objective, with a reference to Principle 15 of the Rio Declaration on Environment and Development, and provisions on an importing Party's decision-making process regarding the import of an LMO: "Lack of scientific certainty due to insufficient relevant scientific information and knowledge regarding the extent of the potential adverse effects of a living modified organism on the conservation and sustainable use of biological diversity in the Party of import, taking also into account risks to human health, shall not prevent that Party from taking a decision, as appropriate, with regard to the import of that living modified organism in order to avoid or minimize such potential adverse effects."

Both the substantive content of the Protocol's precaution provisions and the preambular "savings clause" make clear that a Party's use of precaution in decision-making must be consistent with the Party's trade and other international obligations.

Trade with Non-Parties

The Protocol states that the "transboundary movement of living modified organisms between Parties and non- Parties shall be consistent with the objective of this Protocol." Therefore, although the Protocol only requires trade between Parties and non-Parties in LMOs to be consistent with the "objective" of the Protocol, we anticipate that, as a practical matter, firms in non-Party countries wishing to export to Parties will need to abide by domestic regulations put in place in the importing Parties for compliance with the Protocol.

THE ROLE OF AGRICULTURAL BIOTECHNOLOGY IN WORLD FOOD AID

Biotechnology has the potential to play a key role in reducing chronic hunger, particularly in sub-Saharan Africa, which missed out on the "Green Revolution" of the 1960s and 1970s, says Bruce Chassy, professor and executive associate director of the Biotechnology Center at the University of Illinois Urbana-Champaign. He urges more public investment in agricultural research, education and training at the local, national and regional levels. Food aid is one of several global mechanisms created to deal with hunger and food insecurity.

The need for food aid around the globe varies from specific responses to acute and episodic shortages to long-term donations of food to abate continuing chronic inability of some regions to become agriculturally self-sufficient. While agricultural biotechnology is not a panacea to food insecurity, it is likely to play a vital role in the delivery of food assistance and reduction of hunger for generations to come.

THE GLOBAL NEED FOR FOOD AID

The U.N. Universal Declaration of Human Rights declares the right of access to food and freedom from hunger as a fundamental right. Although we live in a world of unprecedented prosperity and technological development, 800-850 million people are malnourished. More than 200 million of these are children, many of whom will never reach their full intellectual and physical potential. Another 1-1.5 billion humans have only marginally better access to food and often do not consume balanced diets containing sufficient quantities of all required nutrients.

The majority of this nutritionally at-risk population lives in developing countries. Most, perhaps 75 per cent, live in rural agricultural regions. Most are very poor. There is a well-recognized link between poverty and hunger. In fact, family income is probably the single most important determinant of adequacy of access to food. The World Food Summit in 2002 reaffirmed a commitment made by the international community five years earlier to halve the number of hungry people by the year 2015. That goal will not be met unless agricultural productivity and personal income can be improved in the world's poorest regions.

It is argued by some that eliminating poverty is more important than producing more food since there is more than enough food produced in the world to feed everyone. Economists tell us that there is a surplus of food in the world—or at least a surplus of grain that when tabulated as potential caloric intake could theoretically adequately feed the current global population. But the sad lesson of both recent and ancient history is that adequate food supplies do not reach everyone.

The large number of hungry people proves that. It is pointless to argue whether poor agricultural productivity or extreme poverty is more to blame when people are starving. What is clear is that if the rural poor can produce a surplus of food in a more efficient and sustainable manner, there will be adequate food supplies, increasing income and the opportunity for supporting rural development. While most experts would agree that the only long-term solution to hunger is economic development and the elimination of poverty, people who are food self-sufficient through local or regional agriculture will not go hungry. Unfortunately, neither the required increases in agricultural productivity nor the necessary rural development will happen overnight.

The question then becomes "What do we do in the meanwhile?" The short term solution for the hungry is food aid. But even food aid has become politicized as skeptics have charged that it is simply a way for rich over-producing nations to eliminate the surpluses produced by their heavily subsidized farmers. The skeptics also assert that food aid robs local farmers of markets and makes them hungrier. These arguments ignore the daily reality faced by hundreds of millions of hungry people for whom the immediate alternatives are simple: continued hunger and ultimate starvation or the acceptance of food aid.

ELIMINATING CHRONIC HUNGER: A ROLE FOR BIOTECHNOLOGY

The Green Revolution of the 1960s and 1970s helped India and China and other Asian countries become agriculturally self-sufficient net exporters of food in the last three decades. The increased productivity has been accompanied by increases in personal income and stimulus to national economies. Similarly, through application of new technology, agricultural productivity per hectare has doubled in most developed countries in the same timeframe. The development of new high-productivity agricultural technologies resulted from investment in agricultural research performed in government laboratories, research universities, and non-governmental institutes such as the Consultative Group on International Agricultural

Research (CGIAR) centers scattered around the globe. A crucial element of success has been the deployment of effective systems of outreach education and technology transfer. Research and technology transfer has also taken place in the private sector.

For a variety of complex reasons, improvements in agricultural productivity did not take place in all developing countries. Quite the contrary, some of the least developed countries are now even less able to produce sufficient food. There, the Green Revolution never happened. While civil unrest and political corruption may have contributed greatly to this phenomenon, from an agricultural point of view, the failure lies in the lack of investment in and adoption of new technologies and management practices.

Often this occurred because there was not sufficient attention paid or investment made in research to develop effective local or region-specific strategies and technologies. Sub-Saharan Africa is a region where growth in agricultural production has not kept pace with expanding need. As a whole, the region has some of the poorest and most depleted agricultural soils. Only 4 per cent of the farmed land is irrigated. Significant areas of agricultural land are at risk of becoming desert while in some parts of the region excessive humidity and high temperatures contribute to a high incidence of disease and pests.

Weeds such as Striga stifle yields. Droughts are commonplace in some parts of the region. Outright crop failure is common and poor yields are endemic. There is clearly a need to develop crop varieties and management strategies that are more productive under these conditions. High on the list of desired traits are crops with enhanced resistance to environmental stresses such as drought, temperature and salinity; enhanced resistance to diseases and pests; and improved agronomic properties and yield potential. The heavy reliance on a few staple crops makes biofortification—the boosting of the vitamin and mineral components of foods to enhance the nutritional value—an attractive strategy as well. Recent advances in molecular biology and genomics greatly enhance the plant breeder's capacity to introduce new traits into plants. Commercial applications of agricultural biotechnology have already produced crops such as Btmaize, rice, potatoes, cotton and sweet corn (sweet maize) that can protect themselves against insects; virus-resistant papaya, squash and potatoes; and herbicide-tolerant crops such as wheat, maize, sugar cane, rice, onions and beets that allow more effective weed management.

There is accumulating evidence that these biotech crops can be more productive and profitable for farmers. Major reductions in costs for labour, energy and chemicals have been documented. The crops have also proven to be environmentally-friendly, particularly with regard to biodiversity, reduction of agricultural chemicals in soil and water, and decreased exposure of workers and communities to chemicals.

There is also an emerging international consensus of scientific and regulatory opinion that crops derived through biotechnology are safe to eat as food and feed and beneficial for the environment. These and other promising technologies are now being directed at improving the production and yield of African staple crops: banana, cassava, maize, millets, oil crops, peanut, potato, rice, sorghum, soybean, sweet potato and wheat. Protein-enhanced sweet potatoes and potatoes and carotene-enhanced rice and oilseeds promise to improve the nutritional value of the diet.

Thus, over the long term, agricultural biotechnology promises to play a crucial role in improving agricultural productivity and reducing the environmental impact of agriculture leading to agricultural sustainability and food security in many regions of the world. While it would be foolish to say that agricultural biotechnology alone will solve the world's food problems, it would be equally foolish to assert that food insecurity can be eliminated without agricultural biotechnology. In recent years, there has been a significant change in the organization of agricultural research directed at improving food security.

It is now recognized that research needs to be done at local, national and regional levels in order to address specific agricultural challenges and produce new varieties appropriate to local agriculture and customs. This change is particularly focused on utilizing and expanding local scientific and agricultural human and capital infrastructure that can work in partnership with international scientists and funding. Although the path is clear and there are numerous successful examples of these kinds of international partnership, global funding for such activities falls far short of the level required to achieve global food security in the next decades.

RECENT CHALLENGES POSED BY ACUTE FOOD SHORTAGES

Widespread local or regional crop failure often leads to acute food shortages and hunger. The reason for episodic events can be as varied as flood, droughts or civil war. The United Nations, national governments and an assortment of nongovernmental organizations (NGOs) often respond by mobilizing an immediate food aid programme. Food aid distribution can be hindered by lack of infrastructure for storage and transportation of food, and there are often concerns for the security of aid workers. Recently, a new obstacle to food aid distribution has been identified. Repeated crops failures in Southern Africa have placed millions of people in six nations at risk. In response, the United States offered food aid that included substantial shipments of maize. The maize supply in the United States is approximately 30-35 per cent insect-protected Bt-maize developed through biotechnology. This variety of maize had been approved by the U.S. Environmental Protection Agency (EPA), the U.S. Department of Agriculture (USDA) and the Food and Drug Administration (FDA) as safe for consumption as food and feed.

It was commingled with conventional maize in the U.S. commodity system. However, since the intended recipient nations did not use biotech seed varieties and imported few commodities such as maize, they for the most part lacked specific laws and regulatory systems with respect to foods produced through biotechnology. Genetically modified (GM) maize was an unapproved food in their regulatory systems. In light of the global scare campaign against GM foods, several countries hesitated to accept the aid. Ultimately, intensive international consultation and fact-finding satisfied all of these countries save Zambia, which continued to refuse GM food aid. One obvious conclusion to be drawn from this experience is that regulatory systems and training need to be in place before the need for food aid arises again.

Public Investment in Research, Education and Training

What the experience of recent decades has taught is that agricultural biotechnology can be a powerful tool in the development of improved crop varieties for developing countries. The promised benefits can only be realised in a permanent and sustainable manner when the countries that benefit play a role in defining the need, developing the solution and implementing the education and technology transfer systems. Each nation must decide what agricultural goals are in its national interest and what technologies are consistent with consumer acceptance and customs. Shared ownership leads to good stewardship. Partnerships that lead to shared ownership can solve another challenge to applying technology. One major concern about agricultural biotechnology is that the seeds are owned and sold by large multi-national corporations who might eventually exert external domination and control local seed markets and farmers. An additional problem is that developing countries may have limited access to intellectual property rights that would provide them access to modern agricultural technologies such as new seed types.

To help counter these challenges and promote public sector uses in developing countries, a consortium of public universities and public sector institutions has recently announced the formation of the Public Sector Intellectual Property Resource for Agriculture (PIPRA). PIPRA will work to make public sector research available to more of the people who want it and insure freedom to operate. Multi-national corporations have also demonstrated their willingness to donate their technology and expertise to such efforts. There is a holistic answer to all these food security needs and concerns.

The global community needs to invest more capital in creating agricultural institutions and infrastructure in countries that face food security challenges. Investment must be made in legal and regulatory systems, agricultural research, transportation and processing systems, and education. The success of the Land Grant University system in improving agriculture and contributing broadly to society in the United States over the last 140 years demonstrates that the development of human capital and educational systems is as important as scientific discovery. The creation of institutions and public/foundation funding mechanisms would create a platform for international collaboration that is open to government, university and private-sector collaborators. If the world community is to arrive at its stated goal of food security for every person, it must put aside ideological and political divisions and pragmatically embrace each technology that leads to sustainable food security.

THE ROLE OF PLANT BIOTECHNOLOGY IN THE WORLD'S FOOD SYSTEMS

At the molecular level, different organisms are quite similar, writes Cornell University Professor A.M. Shelton. It is this similarity that allows the transfer of genes of interest to be moved successfully between organisms, therefore, genetic engineering is a much more powerful tool than traditional breeding in improving crop yields and promoting environmentally friendly production methods. For the past 10,000 years, humans have used the plants nature provided and modified them through selective breeding to have desirable characteristics such as improved taste, enhanced yield and pest resistance.

The result is that the plants we consume today would be largely unrecognizable to our ancient ancestors. Scientists consider the techniques of biotechnology to be an aid in the selective breeding of plants and to have far more potential for providing benefits such as enhanced nutritional properties, more environmentally friendly production methods and improved yields. Already, the techniques of biotechnology have produced tremendous benefits in medicine. Virtually all the insulin used to treat diabetes today is produced through biotechnology and genetic engineering, and many of the medicines used to fight cancers and heart problems are produced through these same methods.

DEVELOPMENT OF PLANT BIOTECHNOLOGY

Corn (maize) originated in Mexico from a grass called teosinte that has a small reproductive structure bearing little resemblance to the ear of corn seen in markets around the world today. Tomatoes and potatoes first appeared in South America - tomatoes as small fruits the size of a grape, and potatoes as knobby tubers with high concentrations of a family of bitter chemicals called glycoalkaloids, which are toxic to humans.

Through selective breeding by our ancestors, the shape, colour and chemical content of these and hundreds of other plants consumed today have been modified to suit consumer preference or to obtain desired characteristics such as high yield, disease and insect resistance, and tolerance to drought and other plant stresses. Not only have these plants changed in appearance and composition, they also have become distributed worldwide through centuries of human migration and trade. For example, cabbage, which originated in Europe, is now grown on every inhabited continent. When today's consumers walk into a market in many parts of the world, they are witnesses to today's global food system where foods produced in one part of the world are daily shipped to local markets.

We now realise our ancient ancestors were modifying the genetic makeup of plants by transferring genetic material from one plant to another. However, it wasn't until Gregor Mendel, an Austrian monk, conducted experiments in the 1800s with garden peas that the basic laws of heredity were first unraveled. Prior to the early 1900s, traditional plant breeding, like that practiced by Mendel, relied on man-made artificial crosses in which pollen from one plant species was transferred to another sexually compatible plant.

The goal was to take a desirable trait from one plant and introduce it into another plant. However, often desirable characteristics either were not present in sexually compatible plants or did not occur in any plant species. This led plant breeders to seek new ways of transferring desirable genes. Beginning in the 1930s, plant breeders developed techniques to allow them to develop plants from two parent plants that could not normally produce viable offspring. An example is the technique called "embryo rescue," in which the new plant embryo is provided with extra care in the laboratory to enable it to survive during its early growth.

In the 1950s, plant breeders also developed methods of creating variation in an organism's genetic structure through what is termed "mutation breeding." Mutations in the genetic makeup of a plant occur continuously and randomly in nature through such events as the sun's radiation and may lead to the occurrence of new desirable traits. Mutation breeding uses similar random processes to cause changes in a plant's genes. Plants then are assessed to determine if the genes were changed and whether the changes provided a beneficial trait such as disease or insect resistance. If the plant was "improved," then it was tested for other changes that may have occurred. Many of the common food crops we use daily have been developed through techniques such as embryo rescue and mutation breeding and virtually all the foods we consume have genes in them.

It is hard to think of an example of a common food crop in the developed world that has not been improved by some form of modern technology, or what can be termed "biotechnology." Simply put, biotechnology is a set of techniques that utilizes living organisms, or parts of organisms, to make or modify products, improve plants or animals, or develop microorganisms for specific purposes. This definition encompasses all human activities conducted on living organisms from the earliest development of plant breeding 10,000 years ago to the present. This is the reason plant breeders consider the term "genetically modified organisms" - or GMOs - to be misnomer since all common food crops of today have been so modified.

THE SCIENCE OF MODERN GENETIC ENGINEERING

Genetic engineering is one form of biotechnology and usually refers to copying a gene from one living organism—plant, animal or microbe—and adding it to another organism. In genetic engineering, a small piece of genetic material (DNA) is inserted into another organism to produce a desired effect. This is in contrast to traditional plant breeding in which all the genes desirable and undesirable contained in the male plant—pollen—are combined with all the genes of the female plant.

The progeny resulting from this cross may contain the gene for the desirable character, but it will also contain many of the undesirable genes from both parents. Genetic engineering has the advantage of being able to transfer only the gene of interest and greatly accelerate plant breeding. But genetic engineering also is more powerful than traditional breeding since it can move genes not only between similar plant species but also from distant relatives, including non-plant species. It is possible to move genes between such seemingly unrelated organisms because all living organisms share the same code for DNA and the synthesis of proteins and other basic life functions. What might seem on the surface to be very different organisms are, in fact, very similar, at least at the molecular level.

All living things are more alike than different, and this is one of the reasons that genes can be moved so successfully between such seemingly different organisms as plants and bacteria. Genes are not unique to the organisms from which they came, so there really aren't "tomato genes" or "bacterial genes." It's the collection of all genes in a tomato or a bacterium that makes it a tomato or bacterium, not a single gene. As we learn more about the genetic makeup of all organisms, we see that most plant species differ by only a small percentage of their genes and those even such seemingly different organisms as tomatoes and bacteria have many common genes.

These findings suggest that in the long-term evolutionary process even tomatoes and bacteria had some common ancestor. From the discovery 50 years ago of the structure of DNA, scientists soon came to realise they could take segments of DNA that carried information for specific traits—genes—and move them into another organism. In 1972, the collaboration of Hubert Boyer and Stanley Cohen resulted in the first isolation and transfer of a gene from one organism to a single-celled bacterium where it would express the gene and manufacture a protein.

Their discoveries led to the first direct use of biotechnology—the production of synthetic insulin to treat people with diabetes—and the start of what is often called modern biotechnology. Plants were first transformed through genetic engineering in the late 1970s. Mary-Dell Chilton and colleagues used a common soil-dwelling bacterium, Agrobacterium tumefaciens, that attaches itself to plants and transfers some of its DNA into the plant. Chilton and her colleagues added a gene to this bacterium, which in turn transferred the gene into a plant where it became part of the plant's DNA. This bacterium is still commonly used in genetic engineering along with another technique that uses a high-velocity mechanism to inject DNA into plant cells. The result from either technique is the same—the plant cells take up the gene and begin to express it as their own.

Benefits and Risks of Biotechnology

Plants developed through genetic engineering were first grown on 1.7 million hectares in 1996 in the United States, but by 2002 they were grown on 58.7 million hectares in 16 countries. By far the major use of the present plants is to manage pests—weeds, insects and diseases. Weed management with genetically engineered plants is accomplished because the plants have a modified enzyme (a protein) that allows them to survive an application of a specific herbicide that normally acts on that enzyme. Growers can plant the herbicide-tolerant seeds, allow the plants to emerge along with any weeds in the field and then treat the field with an herbicide.

The result is that the weeds, but not the crops, die. The advantage to growers is that they spend less time on weed management, have enhanced weed control, use safer herbicides, and in many cases use less herbicides. Additionally, this technology allows growers to use soil conservation practices such as reduced or no-tillage, thus helping to retain soil structure and moisture and reduce erosion. Herbicide tolerant crops (soybean, canola, cotton and maize) were grown on 48.6 million hectares in 2002. Insect-resistant crops developed through genetic engineering utilize the common soil bacterium, Bacillus thuringiensis (Bt), which has been commercially used for more than 50 years, as an insecticide spray.

Although safe to humans and the environment, when a susceptible insect ingests Bt, the Bt protein binds to specific molecular receptors in the gut and creates pores causing the insect to starve to death. Insecticidal products containing Bt were first commercialized in France in the late 1930s, but even in 1999 the total sales of Bt products constituted less than 2 per cent of the total value of all insecticides. Bt, which had limited use as a foliar insecticide, became a major insecticide only when genes that produce Bt toxins were engineered into major crops. The Bt crops available at present are maize and cotton. These were grown on a total of 14.5 million hectares in 2002. Virus-resistant crops were created by inserting a non-infective part of a plant virus into a plant, essentially "vaccinating" the plant to protect it from the virus.

This technique is called "pathogen-derived resistance." Squash and papaya have been engineered to resist infection by some common viruses and are approved for sale in the United States. There are fewer than 1 million hectares of these crops. The bioengineered plants available at present provide growers with better tools to manage pest problems. As with any technology, there are risks and benefits to currently available genetically engineered plants, but the present body of information indicates their use has enhanced pest management, substantially reduced the amounts of pesticides used in some crops, enabled growers to use safer pesticides, and contributed to enhanced safety for humans and the environment.

The regulatory process for managing these plants and their effects on the environment and humans has evolved with the technology and the scientific community's knowledge of these tools. Many of the more controversial issues surrounding genetic engineering of plants—such as pesticide resistance, gene flow and intellectual property issues—are not unique to this new technology but pertain to all types of agriculture. Some species of insects have developed resistance to sprays of Bt, indicating the potential for some species to become resistant to Bt plants.

However, despite Bt plants being grown on more than 62 million hectares worldwide from 1996 to 2002, there have been no documented cases of resistance development. The reasons for this lack of resistance appear to involve not only biological factors of the insects and the Bt plant, but also the fact that the regulatory agency (the Environmental Protection Agency) in the United States requires a resistance management plan for growing Bt plants. No other insecticide has such strict regulations. Still, growers, companies and federal regulatory agencies must be vigilant about resistance developing for biotech crops used to manage insects, weeds and viruses as they also must with non-biotech pest management tactics. It will be important to consider the accrued environmental and health benefits of these biotech crops prior to the development of any resistance and how resistance can be managed if and where it occurs. In addition to pesticide resistance, gene flow from biotech to non-biotech crops may also be a concern. However, the risk of gene flow varies with each crop and each gene. Pollen flow in soybeans is very limited so the risk of a biotech soybean crop crossing with a non-biotech soybean is minimal, but this may be different for another crop. Likewise, if the gene in the biotech crop that provided a pest management trait, such as insect resistance, moved into a non-biotech plant, such as a weed, any selective advantage of the insect-protected weed in the ecosystem should be assessed. These same questions should also be answered with non-biotech crops, but these have not received the same level of attention as biotech crops because of the latter's higher profile.

Horizon of Biotechnology

In the future, the potential uses of plant biotechnology are far more wide-ranging than the pest-management biotech crops of today. Plants are being developed that serve as production "factories" for medically important drugs, sources of alternative energy, tools for cleaning toxic waste sites, and biomaterials including dyes, inks, detergents, adhesives, lubricants, plastics and the like. Consumers may see these products as more directly enhancing their quality of life than the pest-management biotech crops of today. Perhaps an even more dramatic advantage to consumers will be seen when plants are genetically engineered to have enhanced health benefits such as disease-fighting chemicals or increased amounts of essential vitamins and minerals. A healthy and well-informed discussion of the risks and benefits involved in agricultural biotechnology is needed to ensure a proper role for this new technology in our future food and health systems. No one should believe that any technology, including biotechnology, will completely solve the world's agricultural problems. Many people familiar with biotechnology, however, believe it to be an important component of the solution.

IMPROVING ANIMAL AGRICULTURE THROUGH BIOTECHNOLOGY

Livestock feed derived from biotechnology has been shown to increase production efficiency, decrease animal waste and lower the toxins that can cause sickness in animals, asserts Terry Etherton, distinguished professor at The Pennsylvania State University. Genetically modified feed also can improve water and soil quality by reducing levels of phosphorous and nitrogen in animal waste. Over the past 20 years, biotechnology has lead to the development of new processes and products that have benefited agriculture and society.

Between 1996 and 2002 there was a 35-fold increase in acreage planted globally with genetically modified (GM) crops, from 1.7 to 58.1 million hectares, and more than a quarter of GM crops are grown in developing countries. While there has been considerable discussion about the benefits of GM crops in the grains and fruits humans consume, less public debate has been forthcoming about GM crops' profound effects on improving the health of livestock grown for meat products and on reducing some of the environmental costs of livestock wastes.

Adoption of products produced by modern biotechnology will be important to enable the production of food sufficient to feed a growing world population. Biotechnologies that enhance productivity and productive efficiency—feed consumed per unit of milk or meat produced—have been developed and approved for commercial use in many countries. New biotechnology products have enabled improvements to be made that increase food safety and improve animal health. Biotechnology also offers considerable potential to animal agriculture as a means to reduce nutrients and odours from manure and volume of manure produced. Development and adoption of these biotechnologies will contribute to a more sustainable environment.

In order to be approved for commercial use in the United States, new agricultural biotechnologies are evaluated rigorously by the appropriate federal regulatory agencies to ensure efficacy, consumer safety, and animal health and well-being. Successful development and adoption of emerging biotechnologies for agriculture require improved public understanding of scientific, economic, legislative, ethical and social issues. The objective of this paper is to provide a brief overview of some of the existing and emerging modern agricultural biotechnologies that affect animal productivity and discuss their current or potential food safety and environmental benefits.

feeding livestock

Scientific studies evaluating feed components derived from genetically modified (GM) plants have focused on beef cattle, swine, sheep, fish, lactating dairy cows, and broiler and layer chickens, and have included nutrient composition assessments, digestibility determinations and animal performance measurements. These studies have shown that feed components derived from GM plants are equivalent in terms of nutrient composition to non-GM plants. Feeding components derived from GM plants, such as grain, silage and hay, also show results in growth rates and milk yields that are equivalent to those food components derived from non-genetically enhanced feed sources. Studies have reported that GM maize altered for protection against the "corn borer" can have lower contamination by mycotoxins—toxic substances produced by fungi or molds—under certain growing conditions, resulting in safer feed for livestock. In the United States, there is a long history of assessing the safety of foods introduced into the marketplace. The assessment of GM plants and animal biotechnologies is science-based and rigorous. The discovery and development of ne

In the United States, there is a long history of assessing the safety of foods introduced into the marketplace. The assessment of GM plants and animal biotechnologies is science-based and rigorous. The discovery and development of new animal and plant biotechnologies are part of a continuum leading to the commercialization of agricultural biotechnology products.

Historically, equivalence of composition GM plants, GM animals or animals treated with biotechnology products, such as bST, has been an important component of the regulatory process. Establishing equivalence of composition is evidence that substantive changes did not occur in the plant or animal as the result of the genetic modification event. One endorsement of the robust nature of the comparative safety assessment process used with GM plants is that more than 223 million hectares of GM crops have been commercially grown over the past 10 years with no documented effects to humans, animals or the environment.

Likewise, there have been no documented adverse effects of meat and milk derived from cows supplemented with bST, the most rapidly adopted animal biotechnology to date. Agriculture is transiting a remarkable scientific era with respect to the myriad of processes and products that have been developed using biotechnology. Moreover, many new products of biotechnology are being developed that will benefit the food sector.

Implicit to approval of these new products is a robust safety assessment process. To date, the approved GM plants and animal biotechnologies have been judged to be as safe as conventionally produced counterparts. Development and adoption of new biotechnologies will be crucial in meeting the challenge of producing enough food for a growing world population while reducing impacts on the environment.

The impact these technologies have on society in the future, however, will be largely dependent on the extent to which they are adopted by producers and the agricultural community and accepted by consumers. Questions about societal impacts and safety often arise as the result of technological change. Inherent to the successful development and adoption of new biotechnologies for agriculture is the need to increase public understanding of the scientific, economic, legislative, ethical and social issues associated with emerging agricultural biotechnologies.

BIOTECHNOLOGY IN THE GLOBAL COMMUNICATION ECOLOGY

Public debates about the safety of new products introduced in the market go back centuries and were often based less on science than on the politics of the time. Similarly, today, much of the debate about agricultural biotechnology is steered by myths and misinformation and not by science, writes Calestous Juma, professor and director of the Science, Technology and Globalization Project at the Kennedy School of Government, Harvard University. The scientific community, with stronger support from governments, must do more to address science and technology issues with the public, he adds. Debates over biotechnology are part of a long history of social discourse over new products. Claims about the promise of new technology are at times greeted with skepticism, vilification or outright opposition—often dominated by slander, innuendo and misinformation.

Even some of the most ubiquitous products endured centuries of persecution. For example, in the 1500s Catholic bishops tried to have coffee banned from the Christian world for competing with wine and representing new cultural as well as religious values. Similarily, records show that in Mecca, in 1511 a viceroy and inspector of markets, Khair Beg, outlawed coffeehouses and the consumption of coffee.

He relied on Persian expatriate doctors and local jurists who argued that coffee had the same impact on human health as wine. But the real reasons lay in part in the role of coffeehouses in eroding his authority and offering alternative sources of information on social affairs in his realm. In public smear campaigns similar to those currently directed at biotech products, coffee was rumored to cause impotence and other ills and was either outlawed or its use restricted by leaders in Mecca, Cairo, Istanbul, England, Germany and Sweden. In a spirited 1674 effort to defend the consumption of wine, French doctors argued that when one drinks coffee: "The body becomes a mere shadow of its former self; it goes into a decline, and dwindles away. The heart and guts are so weakened that the drinker suffers delusions, and the body receives such a shock that it is as though it were bewitched."

Butterfly Stories and Other Misinformation Tactics

Today similarly charged stories are told about genetically modified (GM) foods. In addition to claims about the negative impact of GM foods on the environment and human health, there are wild claims that associate GM foods with maladies such as brain cancer and impotence as well as behavioural changes. Some of these rumors are spread at the highest levels of government in developing countries.

The tactics employed in the debates are equally sophisticated. Critics of the technology have used instruments of mass communication to provide the public with information that is carefully designed to highlight the dangers they attribute to biotechnology. Advocates of biotechnology have often been forced to respond to charges against the technology and have only on rare occasions taken the initiative to reach out to the public.

This is particularly important because the general public does not readily understand the technical details of biotechnology products and so new communication approaches are needed. While advocates of biotechnology have often tried to rely on the need for scientific accuracy, critics employ rhetorical methods that are designed to invoke public fear and cast doubt on the motives of the industry.

The critics draw analogies between the "dangers" of biotechnology with the catastrophic consequences of nuclear power or chemical pollution. Indeed, they use terms like "genetic pollution" and "Frankenstein foods". Critics have also relied on the general distrust of large corporations among sections of the global community to make their case. In addition, they have made effective use of incidents, whose risks they have amplified. A muchquoted study by Cornell University researchers indicated that pollen from GM maize producing a Bt toxin killed the larvae of Monarch butterflies. This study was used to dramatize the impact of biotechnology on the environment. Subsequent published peer explanations of the limitations of the study and refutations of the conclusions did not change the original impression created by the critics of biotechnology.

In this case the real environmental issue was not if GM maize killed monarch butterfly larvae or not. The critical question was what impact the maize had on the environment compared to maize grown with chemical pesticides. It is the issue of relative risks that is important; not simply a single event examined outside the wider ecological context. But apparently, this kind of analysis would not serve the cause of critics.

It is notable that the critics of biotechnology have defined the rules of the debate in two fundamental ways. First, they have managed to create the impression that the onus of demonstrating safety lies with advocates of biotechnology and not on its critics. In other words, biotechnology products are considered unsafe until proven otherwise. Second, they have been effective in framing the debate in environmental, human health and ethical terms, thereby masking the underlying international trade considerations. By doing so, they have managed to rally a much wider constituency of activists who are genuinely concerned about environmental protection, consumer safety and ethical social values.

There is a general view that concerted efforts to promote public debate will improve communication and lead to the acceptance of biotechnology products. This may be the case in some situations. But generally, the concerns are largely material and cannot be resolved through public debate alone. This is mainly because the root causes of the debate lie in the socio-economic implications of the technology and not mere rhetorical considerations. It is possible that public debates will only help to clarify or amplify points of divergence and do little to address fundamental economic and trade issues. What then can be done under the circumstances, especially in relation to developing countries that are currently the target of much of the attention of advocates and critics of biotechnology? Operating in the new global communication ecology will require greater diversity of biotechnology products, an increase in the number of institutional players, enhanced policy research on life sciences and society, and stronger policy leadership.

Products Speak Louder than Words

Much of the debate on the role of biotechnology in developing countries is based on hypothetical claims with no real products in the hands of producers or consumers. Under such circumstances, communication and dialogue are not enough until there is a practical reference point. In other words, rebutting the claim of critics is not as important as presenting the benefits of real products in the market place.

This can best be achieved through collaborative efforts among local scientists, entrepreneurs, policy makers and legitimate civil society organizations. There is ample evidence to suggest that concerns over the safety of new products tend to decline as local participation and ownership in new technologies increase. Similarly, local participation in new technologies increases the level of trust in new technologies, thereby reducing the demand for non-science-based safety regulations. For example, the word of a farmer from South Africa stating the positive impact of GM cotton on her welfare carries more weight than thousands of screaming press releases and empty headlines on both sides of the debate. This means that spreading the use of biotechnology not only promotes familiarity with the technology, but also generates the information needed to convince the public about the relevance and usefulness of the technology. The broadening of the range of products is therefore a key aspect of the debate. This is particularly important in developing countries interested in using the technology to enhance local products and diversify their food base. Information on the development of drought-tolerant crops, for example, would be relevant to African countries while other regions might be interested in different products. This view also suggests that general debates about the role of biotechnology are of little utility unless framed in the context of local needs and applications. The absence of a real stake in the technology creates a vacuum that is often filled with misinformation on the risks and benefits of the technology. Countries such as Kenya and South Africa that have their own biotechnology research programs have a more considered view of the technology.

Broadening the Constituency

Addressing the issue of biotechnology communication requires an improved understanding of the changing ecology of communication. The ecology includes a complex network of sources of information and opinion leaders as well as new communication tools that were hitherto not available to the public or advocacy groups. In his days, Khair Beg was outraged to learn that coffeehouses had become an authoritative source of information on what was happening in his jurisdiction. Similarly, the Internet has become a more important communication tool than classical methods such as TV advertising. But unlike in the days of Khair Beg, the new communication ecology is global in character, making it possible to spread information widely and generate empathy among a diversity of activist organizations, including those that are unlikely to be affected by the technology.

These cyber-communities are built around a complex set of mailing lists that are not easily accessible. Correcting misinformation spread through such channels is difficult to do because of the complexity of the networks. While activists tend to use a diverse array of social movements to advance their cause, advocates have tended to focus on the use of centralized institutions whose impact is largely negligible in the modern communication ecology. But creating the necessary diversity requires a broadening of the base of social movements that champion the role of science and technology in human welfare. One of the most important aspects of the biotechnology debate has been the role of the popular media. In Europe, for example, the media have played an important role in amplifying claims by critics or creating doubt about positions advanced by advocates of the technology. In contrast, support for the role of science does not usually have the polemical turn that newspaper editors relish.

The traditional view that science is based on immutable facts which can be passed on from an authority to the general public is being challenged by approaches that demand greater participation in decision-making. In other words, scientific information is being subjected to democratic practices. The debate over biotechnology has pushed the frontiers of public discourse of technical matters. On the one hand, society is being forced to address issues that are inherently technical, and on the other, the scientific community is under pressure to accept non-technical matters as valid inputs to decision-making.

Thinking Ahead

Policy-oriented research institutions and think tanks play an important role in the war of words. It is notable that critics of biotechnology have made a considerable effort to create alliances with research institutions, including university-based departments. Much of the material used to question the safety of biotechnology often has the legitimacy of a research institution. But non-partisan policy research on the role of biotechnology in society is largely lacking, and so those seeking to provide an alternative view have limited opportunities to obtain credible information. The lack of systematic research on the interactions between biology and society is a critical bottleneck in efforts to engage the public in dialogue on biotechnology. This is particularly critical given the fact that advances in biology pose new ecological and ethical issues that are not associated with the physical and chemical sciences. For example, concerns over the inability to recall products once released on markets are more pronounced when dealing with the release of biological inventions into the environment.

Leading the Way

Much of the public debate is intended to influence government policy on biotechnology. In this regard, the capacity of governments to assess the available information and use it for decision-making is an essential element of the debate. Political leadership on biotechnology and the existence of requisite institutions of science and technology advice are an essential aspect of the governance of new technologies. Debates over new technologies will be more pronounced in the future, and governments will increasingly come under pressure to address these issues.

But science and technology advice will not be sufficient unless governments view science and technology as integral to the development process. In this regard, enhancing the capacity of leadership to address science and technology issues will contribute to the effective management of public debates over new technologies in general and biotechnology in particular. On the whole, the nature of emerging technologies—particularly those based on the life sciences—and the changing ecology of communication are making it necessary to rethink strategies for advancing the role of biotechnology in society. The scientific community will need not only to demonstrate a clear sense of leadership, but also to adapt its communication methods to suit the growing complexity and diverse needs of the global community. In the final analysis, it is the range of useful products available to humanity from biotechnology that will settle the debate, not the hollow pronouncements of advocates and critics.

10

Biodiversity and Agricultural Biotechnology

Biological diversity is an imprecise term that may refer to diversity in a gene, species, community of species, or ecosystem. It is often contracted to *biodiversity* and used broadly with reference to the total biological diversity in an area or the earth as a whole. Biodiversity comprises all living beings, from the most primitive forms of viruses to the most sophisticated and highly evolved animals and plants. According to the Convention on Biological Diversity, biodiversity means "the variability among living organisms from all sources including, *inter alia,* terrestrial, marine, and other aquatic ecosystems and the ecological complexes of which they are part".

Genetic, Species and Ecosystem Diversity

Genetic diversity

Genes are the basic building blocks of life. In many instances genetic sequences, functions and the proteins encoded by the genes are highly conserved across all species. The importance of genetic diversity is noted in the combination of genes within an organism (the genome), the variability in phenotype that they produce as well as their resilience and survival under selection. As such, it is widely believed that ecosystems, natural ecosystems in particular, should be managed in a way that protects the untapped resource of genes within their host organisms. Today, much work remains to be done to both characterize genetic diversity and understand how best to protect and make wise use of it.

Species diversity

For most practical purposes, species are the most appropriate targets for biodiversity research and species diversity is the most useful measure of biodiversity. There is no single definition of what a species is and species-level taxonomy can change with new data as well as new approaches. Nevertheless, a species could broadly be defined as a collection of populations that differ genetically from one another to a greater or lesser degree.

These genetic differences manifest themselves as differences in morphology, physiology, behaviour and life histories; in other words, genetic characteristics affect expressed characteristics (phenotype). Today, about 1.75 million species have been described and named but the majority remains unknown. The global total might be ten times greater, many of these being undescribed insects.

Table. Estimated Numbers of Described Species and Possible Global Total

Kingdoms	Phyla	Described Species	Estimated Total
Bacteria		4,000	1,000,000
Protoctista		80,000	600,000
Animalia			
	Craniata (vertebrates) total	52,000	55,000
	Mammals	4,630	
	Birds	9,946	
	Reptiles	7,400	
	Amphibians	4,950	
	Fishes	25,000	
	Mandibulata (insects & myriapods)	963,000	8,000,000
	Chelicerata (arachnids etc)	75,000	750,000
	Mollusca	70,000	200,000
	Crustacea	40,000	150,000
	Nematoda	25,000	400,000
Fungi		72,000	1,500,000
Plantae		270,000	320,000
TOTAL		1,750,000	14,000,000

Ecosystem Diversity

At its highest level of organization, biodiversity is characterized as ecosystem diversity, which can be classified in the following three categories:

- Natural ecosystems, *i.e.* ecosystems free of anthropogenic management activities. These are composed of what has been broadly defined as "Native Biodiversity". It is a matter of debate whether any truly natural ecosystem exists today since man has influenced with its activity most regions on earth.

- Semi-natural ecosystems, in which human activity is limited. These are important ecosystems that are subject to some level of low intensity human disturbance. These areas typically abut managed ecosystems.
- The third broad classification of ecosystems is "managed". Such system can be managed to varying degrees of intensity from the most intensive, conventional agriculture and urbanized areas, to less intensive systems including some forms of agriculture in emerging economies.

Beyond simple models of how ecosystems appear to operate, we remain largely ignorant of how they function, how different ecosystems might interact with each other, and which ecosystems are critical to the services most vital to life on Earth.

The role of the forests for water management is crucial in all forested habitats, from the temperate to the tropical rainforests, and due to acute threats through urbanisation, in particular also the dry tropical forests. Because we know so little about the ecosystems that provide our life-support, we should be cautious and work to preserve the broadest possible range of ecosystems.

Distribution of Biodiversity

Biodiversity is not distributed evenly over the planet. Species richness is highest in warmer, wetter, topographically varied, less seasonal and lower elevation areas. There are far more species in total per unit area in temperate regions than in polar ones, and far more again in the tropics than in temperate regions. Latin America, the Caribbean, Asia and the Pacific host together 80% of the ecological megadiversity of the world.

Within each region, every specific type of ecosystem will support its own unique suite of species, with their diverse genotypes and phenotypes. In numerical terms, global species diversity is concentrated in tropical rain forests. The Amazon basin contains for example 87 to nearly 300 different tree species per hectare and supports the richest fish fauna known, with more than 2500 species. The forests in Asia and South America are considered to be especially rich in animal species.

Species and genetic diversity within any agricultural field will be more limited than in a natural or semi-natural ecosystem. Nevertheless, agricultural ecosystem can be dynamic in terms of species diversity over time due to the amount of management. Biodiversity in agricultural settings is extremely important at country level in areas where the proportion of land allocated to agriculture is high. This is the case in Europe for example, where 45% of the land is dedicated to arable and permanent crop or permanent pasture. In the UK, this figure is even higher, at 70%.

Loss of Biodiversity

Threats to Global Biodiversity

Loss of biodiversity is occurring in many parts of the globe, often at a rapid pace. It can be measured by loss of individual species, groups of species or decreases in numbers of individual organisms. In a given location, the loss will often reflect the degradation or destruction of a whole ecosystem. Recently the Subsidiary Body on Scientific, Technical and Technological Advice of the Convention on Biological Diversity ranked threats to global biodiversity in the following manner:

- Habitat loss
- Introduction of exotic species
- *Further*: Flooding, lack of water, climate changes, salination etc., all of which may be either natural or man-made (not dealt with in this report).

Today, more than half of humankind lives in urban areas, a figure predicted to increase to 60% by 2020 when Europe, Latin America and North America will have more than 80% of their population living in urban zones. Five thousand years ago, the amount of agricultural land in the world is believed to have been negligible. In 2000, arable and permanent cropland covered approximately 1,497 Smillion hectares of land, with some 3,477 million hectares of additional land classed as permanent pasture. The sum represents approximately 38% of total available land surface (13,062 million ha).

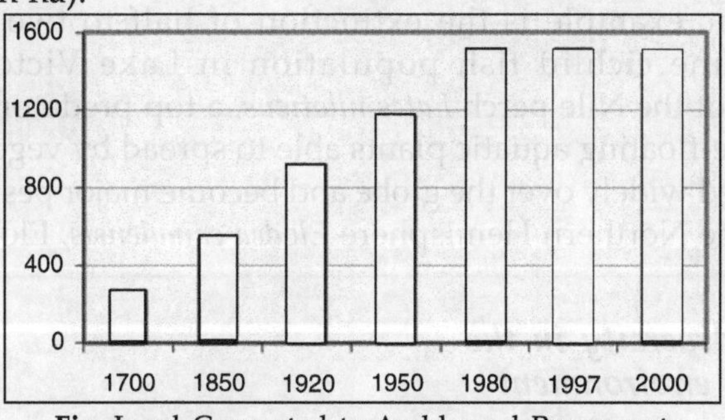

Fig. Land Converted to Arable and Permanent Cropland, in Million Hectares.

Habitat loss is of particular importance in regions of high biological diversity where at the same time food security and poverty alleviation are key priorities (*e.g.* some parts of Latin America and Asia Pacific for example). Forests are a good example: the impacts of development activities and the advance of the agricultural frontier has led to an overall decline in the world's forests and woodlands of approximately 2% between

1980 and 1990. While the area of forest in industrialised regions remained fairly unchanged, natural forest cover declined by 8% in developing regions.

Introduction of exotic species

Unplanned or poorly planned introduction of non-native species and genetic stocks is a major threat to terrestrial and aquatic biodiversity worldwide. There are hundreds if not thousands of new and foreign genomes introduced with trees, shrubs, herbs, microbes and higher and lower animals each year. Many of those survive and can, after years and even many decades of adaptation, begin to be invasive.

Terrestrial areas most affected by the introduction of exotic species include forests, Mediterranean regions as well as similar types of natural vegetation in the Cape Province of South Africa, parts of Chile, Southern Australia and California, grasslands and savannas and agricultural lands. One of the most extreme examples is seen in the pampas of Argentina, a flat grassland with a moderate climate, from which nearly all the native grasses have disappeared and have been replaced by European plants. Islands and other areas having evolved unique ecosystems are particularly at risk.

Freshwater habitats worldwide are amongst the most modified by humans, especially in temperate regions. In most areas, introduction of non-native species is the most or second most important activity affecting inland aquatic areas, with significant and often irreversible impacts on biodiversity and ecosystem function.

A classic example is the extinction of half to two thirds of the haplochromine cichlid fish population in Lake Victoria after the introduction of the Nile perch *Lates niloticus,* a top predator. Also, several species of free-floating aquatic plants able to spread by vegetative growth have dispersed widely over the globe and become major pests, as a notable example in the Northern Hemisphere *Elodea canadensis,* Elodea, Common Waterweed.

Loss of biodiversity in the agricultural environment

In an agricultural context, a rapid decline in species and genetic diversity has been brought about by the success of new commercial varieties. Reported losses of over 80% of varieties in species such as apple, maize, tomato, wheat and cabbage have occurred worldwide. Studies in population genetics raised concern over genetic erosion and the recognition of the importance of plant genetic material in the development of new varieties led to the establishment in the 1970's of the International Plant

Genetic Resources Institute and increased efforts to collect germplasm for *ex-situ* collections.Terrestrial but also aquatic biodiversity within and around agricultural fields, as discussed in Chapter 3, has also been strongly influenced by agricultural practices. Fertilizers, pest control chemicals, tillage and even crop rotation have been shown to profoundly impact the richness and diversity of agricultural ecosystems.

Importance of Biodiversity

Biological diversity has emerged in the past decade as a key area of concern for sustainable development. It provides a source of significant economic, aesthetic, health and cultural benefits. The well-being and prosperity of earth's ecological balance as well as human society directly depend on the extent and status of biological diversity.

Biodiversity plays a crucial role in all the major biogeochemical cycles of the planet. Plant and animal diversity ensure a constant and varied source of food, medicine and raw material of all sorts for human populations. In agriculture, biodiversity represents a critical source of genetic material allowing the development of new and improved crop varieties.

In addition to these direct-use benefits, there are enormous other less tangible benefits to be derived from natural ecosystems and their components. These include the values attached to the persistence, locally or globally, of natural landscapes and wildlife, values, which increase as such landscapes and wildlife become scarcer.

Ever since the first Stockholm Report of the United Nations Environmental Programme, biodiversity has been indirectly on the global agenda.

In view of the importance of biodiversity for the future of mankind, several international agreements aiming at relieving some of the pressure on selected important resources have been reached.These include for example the numerous regional fishery management schemes, the Convention on International Trade in Endangered Species (CITES), and more recently the Convention on Biological Diversity.

Agricultural Practices

A number of tools are currently in use to produce sufficient food for the world. This was accomplished with only a 1.1 fold increase in cultivated lands and was made possible due to dramatic changes in agricultural practices including use of fertilizers and pest control compounds, implementation of specific agricultural practices, shifts to higher yielding varieties and adoption of new technologies. The following section will review current common agricultural practices that are used to increase productivity.

Agricultural inputs

The productivity of crop plants is challenged by abiotic and biotic stresses. Abiotic stresses include nutrient deficiencies, water challenges as well as soil acidity, alkalinity and salinity. Fertilization and irrigation are two important tools for addressing these problems. Nitrogen and phosphorous fertilizers are commonly applied. In fact, the doubling in world food production cited above was accompanied by a 6.87 fold increase in nitrogen fertilization and a 3.48 fold increase in phosphorous fertilization.

In that same time, water challenges were met by increasing irrigated lands by 1.68 fold. Biotic stresses include weeds, insects and plant pathogens such as fungi, viruses and bacteria. A number of pesticides are commonly used to control these pests. Nevertheless, between 35-42% of the world's food and fibre is lost from damage by pests despite the use of 2.5 million metric tons of pesticides;. Weeds cause 10-13% loss, insects 13-16% and pathogens 12-13%. Without pesticides or other pest control measures, it has been estimated that the losses would increase to 70% with an economic loss of $400 billion USD per year. Pest control measures are a positive economic investment for farmers yielding a return of $3-4 USD for each dollar invested.

Weeds are a major problem in many crops so herbicides are an important tools in these crops. Over 90% of US soybean acreas and 70% of Brazilian and Argentine soybean acres are treated with herbicides. For maize, over 95% of US acres are treated with herbicides. Herbicide tolerant crops can provide an opportunity to reduce herbicides in such systems.

In the US, an average 10% reduction in herbicide usage was seen with herbicide tolerant soya from 1995-1998. A more recent study found a reduction of 28.7 million pounds of active ingredient in herbicide tolerant soya in the US in 2001. In the EU, a standard maize herbicide programme uses approx-imately 1740 g of active ingredient per hectare but this amount could be reduced by 30-60% if GM technology were adopted. Similar levels of herbicide reductions were projected for winter oil seed rape (for UK) and sugar beet (for Denmark) of GM crops with herbicide tolerance were adopted in these countries. Control of other pests is critical in a number of crops. High levels of insecticide are used to control ravaging insects in many of the world's cotton growing areas for example, reducing crop losses in some regions from 35—39% to 13% or less. Adoption of insect protected cotton has impacted the level of insecticides used this crop in many areas.

In the US, the estimated savings in metric tons (MT) of active ingredient are 848 in 2001, 1224 MT in 1999 and 907 MT in 1998. In China, an 80% reduction in kg of formulated product used was seen due to the adoption on GM cotton. Introducing GM cotton in Spain would lead to a 60% reduction in volume of pesticide used and nearly a 40% reduction in active ingredient used.

Cultural Practices

Crop Rotation

Crop rotation is a very common practice as a means of controlling pests. Since some pest species rely on specific crops ass hosts, then rotating to another crop can reduce populations of such pests. Crop rotation has been applied in virtually all agricultural strategies, from classic and historic agriculture such as the one still in place in certain actively protected localities in the Swiss Valais. The maize/soybean rotation in the United States as a means of controlling corn rootworm is one example of such a rotation designed to aid in pest control efforts. Another example is that of glyphosate-tolerant Roundup Ready soybeans which are often rotated with such crops as corn, winter wheat, spring cereals and dry beans. An interesting study from Canada shows enhancement of some agricultural parameters after 8 years in the second rotation cycle: N fertilizer requirement decreased, and wheat yield was 22% higher, under no tillage conditions as compared to conventional tillage.

Tillage

The soil in a given geographical area has played an important role in determining agricultural practices since the time of the origin of agriculture in the Fertile Crescent of the Middle East. Soil is a precious and finite resource. Soil composition, texture, nutrient levels, acidity, alkalinity and salinity are all determinants of productivity. Agricultural practices can lead to soil degradation and the loss in the ability of a soil to produce crops. Examples of soil degradation include erosion, salinization, nutrient loss and biological deterioration. It has been estimated that 67% of the world's agricultural soils have been degraded.

At one time, tillage of soil was an essential tool for the control of weeds. Unfortunately, tillage practices can lead to soil degradation by causing erosion, reducing soil quality and harming biological diversity. Tillage systems can be classified according to how much crop residue is left on the soil surface. Conservation tillage is defined as "any tillage and planting system that covers more than 30% of the soil surface with crop residue, after planting, to reduce soil erosion by water". The value of reducing tillage was long recognized but the level of weed control a farmer required was viewed as a deterrent for adopting conservation tillage. Once effective herbicides were introduced in the latter half of the 20th century, farmers were able to reduce their dependence on tillage. The development of crop varieties tolerant to herbicides has provided new tools and practices for controlling weeds and has accelerated the adoption of conservation tillage practices. Herbicide tolerant cotton has been rapidly adopted since its introduction. In the US, 80% of growers are making fewer tillage passes and 75% are leaving more crop residue. In a farmer survey, seventy-one per cent of the growers responded that herbicide tolerant cotton had the greatest impact on towards the adoption of reduced tillage or no-till practices.

In soybean, the growers of glyphosate tolerant soybean plant higher percentage of their acreage using no-till or reduced tillage practices than growers of conventional soybeans. Fifty-eight per cent of gyphosate-tolerant soybean adopters reported making fewer tillage passes versus five years ago compared to only 20% of non-glyphosate tolerant soybean users. Fifty four per cent of growers cited the introduction of glyphosate tolerant soybeans as the factor which had the greatest impact towards the adoption of reduced tillage or no-till.

Germplasm

Crops and varieties

In agriculture, 7'000 species of plants are used by farmers somewhere in the world, but only 30 species provide 90 per cent of our calorific intake. The top three crops are wheat, rice and maize (corn) occupying 230 million hectares, 151 million hectares and 140 million hectares, respectively, which is 35% of all global cropland. Each of the three major crops originated in different regions of the world. Wheat originated in the Near East, rice in both eastern Asia and western Africa and maize in the Americas.

Within these dominant crop species, there are many hundred thousands of varieties (landraces, cultivars) adapted to local climates, farming practices, and cultural predilections like taste, colour, structure, ability to store the products etc. Much of this large crop diversity is important for providing the initial material for breeding. However, it must be recalled that the genetic diversity found in crops is much less broad than the genetic diversity observed in plants or animals living in the wild, which points to the importance of wild species for agricultural breeding programs.

A major factor in the doubling of food production was the introduction of improved varieties.

For both rice and wheat, these repeated crossings led to varieties with four important characteristics:

1. Higher yield;
2. Fast maturation;

3. Semidwarf growth habitat
4. Resistance to disease.

Once these characteristics were introduced, crossings to local varieties produced crops that are regionally adapted for optimized growth and consumer desires. The improved varieties were not just new seeds, but required the adoption of a suite of new agricultural tools. These tools included inputs such as fertilizers and pesticides, equipment for irrigation and tillage. The new technology package was important to optimize the output of the new varieties and thus realise the tremendous gains in productivity seen with the Green Revolution.

Evolution of plant breeding

For most major crops, breeder's collections are sufficiently large to provide an adequate source of additional genetic material. Material from landraces and con-specific wild populations (primary gene pools) are also frequently called upon. FAO have estimated that 30 - 40% of productivity gains overall have relied on genetic contributions from landraces. The secondary gene pool, consisting of related species in the wild or in cultivation, has also provided important and economically valuable contributions to major crops. However, the difficulty of crossing different species using conventional methods has until now limited the use of this genetic resource. Gene transfer technology has the potential to avoid some of the difficulties limiting conventional techniques and brings the possibility of introducing into cultivars traits from an unlimited gene pool. Such processes could perhaps provide new economic incentives to conserve agricultural biodiversity.

Genetically modified (GM) crops are only the last development stage in a long row of breeding method enhancements, and certainly the list will be longer with the years to come. If one looks closer at the development of modern breeding methods, 13 steps can be discerned before reaching true transformation through genetic engineering, as described in the box below:

13 Steps in Plant Breeding from Mass Selection to Genetic Engineering

1. Selection of characteristic, homogenous varieties from traditional local or indigenous origin (*e.g.* land races) that generally exhibit more or less variable populations, by testing the offspring.
2. Crossing of homogenous varieties to create new variability followed by subsequent selection.
3. Intentional introduction by crossing in of desired traits for instance resistances. (Even if 'genes' such as 'mlo' or 'Lr 27' are in question

the manipulations are carried out with pollen and ears of grain - not with DNA).

4. Artificial infection of plants in greenhouse or field by means of contact with neighbouring infected plants or a concentrated liquid spray of fungal spores. This is done to select for resistance. (On account of the high cost this is not widely practised).
5. Intentional use of the heterosis effect in hybrid breeding. (This often requires preparatory steps: some years of inbreeding if cross-pollinated species are involved, or in the case of self-pollinated species, the artificial production of male sterility - cytoplasmic, chemical or by genetic engineering.)
6. Crossing to introduce characters from more distantly related species. (This often necessitates the cultivation of the crossed embryos in a nutritive medium, since embryos left in the seeds would die owing to incompatibility. This is called 'embryo rescue'.
7. Colchicinising (treating with the toxin of the autumn crocus, Colchicum autumnale), to double the number of chromosomes). In many vegetable and fodder crops this enables a stronger expression of certain traits such as frost resistance. It also permits the crossing of two different species or even families, because it can render fertile the sterile offspring of crosses. The most well known example in practice is Triticale, a new species of grain, the result of crossing wheat (Triticum) and rye (Secale), two different families.
8. Inducing mutations with chemicals or ionising radiations and subsequent selection. This method enjoyed a certain boom 10 to 20 years ago but is not much used nowadays since the mutations are mostly disadvantageous. There exist however, short-strawed strains of wheat that were obtained in this way, where you will discover 548 seed propagated crops which have undergone gamma mutation programmes, some of them undoubtedly have also been introduced in later breeding programmes of all kinds.
9. Anther culture. In this way it is possible, in the case of self-fertile heterozygotes whose progeny in the next generation would otherwise diversify, to 'freeze up' so to speak the haploid chromosome set of the pollen or the ovule. The pollen or the unfertilised ovule must be placed on a special sterile nutrient medium, fused by treating with colchicine and raised to become haploid plants, followed by subsequent colchicinising. Thus with one stroke one obtains a homogenous plant which would otherwise only be achieved by many generations of selection. Anther culture

is established mostly in barley and potatoes. With wheat and maize it is still at the experimental stage.

10. In-vitro-selection. If seedlings or tissue fragments can be selected in culture dishes for resistance against a fungal toxin, the cost of field trials is less because many plants will be discarded from the outset. For many traits, such methods are very successful and great efforts are being made to introduce them into routine breeding.

Selection for traits:

11. Somatic hybridising (*i.e.* non-sexual fusion of two somatic cells). The advantage of this method is that by the fusion of cells with different numbers of chromosomes (for instance different species of Solanum) fertile products of the crossing are obtained at once because diploid cells are being somatically fused. Polyploid plants are obtained containing all the chromosomes of both parents instead of the usual half set of chromosomes from each. For this, cells are required whose cell walls have been digested away by means of enzymes and are only enclosed by a membrane, (these are then called protoplasts). With the loss of their cell walls, protoplasts have also lost their typical shape and are spherical like egg cells. This mixture of cells to be fused is then exposed to electric pulses. In order to get from the cell mixture the 'right' product of the fusion (since fusion of two cells from similar plants can also occur) one different selectable character in each of the original plants is necessary. Only cells that survive this double selection are genuine products of fusion. (The easiest way to achieve such selectable markers is by genetic engineering, for instance by incorporating antibiotic resistance into the original plants.) Protoplast fusion has been investigated and applied to potatoes, for instance. In the EU regulations concerning the deliberate release of genetically modified organisms into the environment somatic hybrids are not considered as GMO's and do not require authorization.25 The most recent draft of the EU organic regulations in which the introduction of GMOs in organic cultivation is forbidden, follows the above definition.

12. Marker-assisted selection. For the purpose of diagnosis, DNA from all the plants from which selection is to be made, is isolated and, with the help of enzymes, broken up into smaller or larger pieces. These, in a gel-like mass, are separated in an electric field (gel electrophoresis) according to their size. By various methods, individual pieces are made visible by radioactive or fluorescent markings. Then, on the gel under ultra-violet light, or on an X-ray

film of the gel, a characteristic band pattern for every genotype can be seen. By means of the enzyme and DNA probe used and from their size (*i.e.* the number of their base pairs determined with the help of standards), these bands can be defined). Presently there are a number of modified methods, but the principle is the same. One looks out for bands that correlate statistically with the particular feature. Once such 'markers' have been found one has a simple criterion for selection. A fragment of leaf of the relevant plant furnishes sufficient DNA for such a diagnosis since the DNA can be artificially multiplied in the laboratory. Selection with the aid of markers is very extensively researched for every sort of cultivated plant and every conceivable trait. At the present time many breeders consider it to be *the* investment for the future that will bring about the greatest changes during the next decade. In the coming years it will be integrated into practically all the major breeding programmes. Above all, it will enormously accelerate the process of breeding. Selection will be automated and take place in the laboratory. It will be possible to reduce field trials drastically. For perennial plants it will be of particular interest as well as for projects where one wants to combine many dominantly inherited resistances to the same disease in one variety (pyramidising). Otherwise there is no means of knowing in such projects whether a further resistance exists if the first is already active. This might provide a contribution to permanent disease resistances. Also for complex traits inherited as polygenes the method would promise a speeding up of selection. This method certainly implies working with isolated DNA, but without invasion of the genome of the plant and is therefore not seriously disputed. One must be aware that much genetic engineering with bacteria was and is necessary to establish marker-assisted selection.

13. Gene transfer. With gene transfer there are also many degrees of departure from the 'natural' according to the origin of the genes and the technology employed in the transfer. Modern breeding will depend more and more on molecular methods, which will develop even faster in the years to come, its success will provoke renewed debates on the exclusion of such methods from certain breeding programmes as *e.g.* in organic farming.

Genetically Modified (GM) Crops

The resulting organism is called a Genetically Modified Organism (GMO). Since all genes consist of DNA, and the information in this DNA molecule is read in the same way in all organisms in order to make proteins, it is in principle possible to take any (single) gene from any organism and transfer it into any other organism so that the recipient produces a protein normally only made in the donor.

From the time this simple strategy was devised and , it took molecular biologists about a decade until the first GM plants were made in 1985. Ten years later, the first GM crop appeared in supermarkets in the USA, the "FlavrSavr" tomato with a delayed ripening process. The FDA's review of the Flavr Savr was requested by the tomato's developer, Calgene Inc. of Davis, California, in August, 1991.

The company later submitted a food additive petition on the use of the kan-r gene in the development of new varieties of tomato, cotton, and rapeseed. In 1990-92, the U.S. Department of Agriculture granted Calgene permission to begin large-scale production of the new tomato, final approval by fax May 1994. Agronomic traits followed in 1996 with the introduction of herbicide tolerant soybean and insect-resistant cotton.

Global Adoption

The adoption of GM crops is, in a worldwide view, a success story without precedent. In 2002, four countries grew 99% of the global transgenic crop area. The USA led the world with 39.0 million hectares (66% of global total) Argentina followed with 13.5 million hectares (23%), Canada 3.5 million hectares (6%) and China 2.1 million hectares (4%). China showed the greatest growth with a 40% increase in its insect resistant cotton area from 1.5 million hectares in 2001 to 2.1 million hectares in 2002.

This represents 51% of the total cotton area of 4.1 million hectares in China. Argentina increased its GM crop area by 14% from 11.8 million hectares in 2001 to 13.5 million hectares in 2002. South Africa increased its growings by 20% to 0.3 million hectares in 2002. The US and Canada both showed a growth rate of 9%. GM cotton area in Australia decreased by half in 2002, due to the very severe drought conditions. India, Colombia and Honduras grew transgenic crops for the first time in 2002. Overall, The number of countries that grew GM crops increased from 13 to 16 in 2002—9 developing countries, 5 industrial and 2 Eastern Europe countries.

Current Products

In 2002, the principal GM crops were: soybean occupying 36.5 million hectares (with 51% of all soybean transgenic), cotton at 6.8 million hectares (12% of all cotton was GM); canola at 3.0 million hectares (12% of canola now GM) and maize at 12.4 million hectares (9% of maize now GM). Herbicide tolerance has consistently been the dominant trait followed by insect resistance. In 2002, herbicide tolerance was deployed in soybean, corn, cotton and canola and occupied 75% or 44.2 million hectares of the global 58.7 million hectares. Herbicide tolerant soybean was the single biggest trait/crop with 36.5 million hectares. Insect protected crops were offered in maize and cotton and covered 10.1 million hectares of the global transgenic area in 2002. Bt maize covered 7.7 million of those hectares. Stacked gene combinations with both herbicide tolerance and insect protected traits in the same product were offered in both cotton and maize and occupied 4.4 million hectares in 2002. A small amount of GM crops with virus resistance was also grown in 2002.

Biotechnology and plant breeding

Biotechnology is a valuable tool in plant breeding from 2 different aspects: as a tool to transfer new genes into crop varieties and introduce desired characteristics (as discussed previously), or as a tool for acquiring knowledge.

Today, biological research can hardly be conducted without using biotechnology in one way or another. Taxonomy and conservation use molecular markers to identify species, much in the same way as in forensic medicine to identify criminals. This is useful for *ex situ and in situ* conservation of plants. In seedbanks and conservation projects, genetic fingerprints are used to establish the origin of a seed or the relatedness of one plant variety to another. There are many texts on the use of molecular biology methods in conservation. Biotechnology also is used for important phylogenetical studies in plant systematics, the application of various methods has led to breakthroughs in systematic botany: Results of the application of modern biological and statistical methods can be seen. It is even possible to use the invaluable collections of herbarium plants in pressed and dryed condition as a good source for DNA studies. Molecular data are surprisingly well matching the non molecular ones, as has been shown by a thorough analysis. Biotechnology has proven useful for following genetic markers in plant breeding. Here, plant varieties are crossed by conventional means. By analysing a few cells of the newly sprouted plant, one can predict some of the expected properties of the progeny, by looking at the presence or absence of certain genes. This enables one to predict a phenotypic property, which will only show up later in life, for instance the crop's expected resistance to an infectious plant disease.

These applications of biotechnology are in general not controversial. Here, the public perception is similar to the application of modern biotechnology in medicine, for instance to industrially produced pharmaceuticals, vaccines and diagnostics. It may be noted that the worldwide sales of biotechnology products in 1999 was worth around 13 billion dollars in the medical field and only 1.6 billion in agriculture.

The availability of genome sequences will be a boost to research. The first two complete plant genome sequences determined were those of *Arabidopsis* and rice. The 120 million base pairs (MBP) of the small brassica *Arabidopsis* were sequenced by an international academic consortium and the data made public. The 430 MBP sequence of rice was completed only a few weeks later by an industrial group lead by Syngenta, and will be available by contract to other researchers. Syngenta intends to make the data available free of charge for research directly benefiting subsistence farmers. The public sector sequencing of rice through an international consortium is expected to be completed in 2004. It will hopefully become common practice for companies to make their basic discoveries publicly available, to everyone's benefit. The Monsanto company has also opened up some of its rice sequencing data. An easy way to follow up the progress is to check the Genomics Gateway.

The Impacts of Agricultural Practices on Biodiversity

The following section discusses the impacts of common agricultural practices on biodiversity, and ways in which some of these impacts can be mitigated. As presented in Section 1, biodiversity can be quantified in several different, equally important ways, thus agricultural impacts on biodiversity are considered both in terms of species and genetic diversity. Within each of these categories, the impacts on agricultural biodiversity and natural biodiversity are addressed separately because the impacts of agriculture are different on these two types of habitat. This distinction could also be thought of as on-site and off-site impacts of agricultural practices.

Impacts on Species Biodiversity

Agricultural Biodiversity

General Impacts of Modern Intensive Agriculture

Modern agricultural practices have been broadly linked to declines in biodiversity in agro-ecosystems. This has been found to be true for a wide variety of taxonomic groups, geographic regions and spatial scales. More specifically, various researchers have found significant correlations between reductions in biodiversity at various taxonomic levels and measures of agricultural intensification.

For example, a review of published studies on arthropod diversity in agricultural landscapes found species biodiversity to be higher in less intensely cultivated habitats. Similarly, analysis of 30 years of monitoring records demonstrated that the abundance of aerial invertebrates at a location in rural Scotland was negatively correlated with a suite of agricultural variables that represent more intensive agriculture; that is, arthropod populations are lowest where agriculture is the most intensive.

In this same study, the abundance of various farmland bird species was, in turn, positively correlated with arthopod abundance in the same year and the previous year. Comparable studies have found similar impacts on bird species throughout the United Kingdom and European Union (EU). Across Europe, declines in farmland bird diversity are correlated with agricultural intensity and declines in the European Union have been greater than in non-Member States. These effects of agricultural intensification undoubtedly reflect a large number of factors which are addressed individually in the following sections, including the cropping pattern, the frequency of tillage, the amount and nature of fertilizers used, and the amount and nature of pesticides applied (particularly insecticides and herbicides). However, it should be kept in mind that all of these factors are interrelated to a greater or lesser degree.

Crop Diversity

Intensive agricultural systems typically have limited crop diversity. Many such systems are monocultures at least at the level of individual fields, and are relatively homogenous even at the regional level.

Low crop diversity generally will mean both limited botanical diversity and limited structural diversity. Robinson and Sutherland analysed changes in agriculture and biodiversity in Britain since the 1940s.

They found a consistent reduction in landscape diversity, as reflected in a 65% decline in the number of farms. Farms had become more specialized and efficient. This also was associated with the removal of 50% of hedgerows and a reduction in winter stubbles. Hedgerows and similar non-cropped habitat are important sources of food and shelter for a variety of plants and invertebrates.

Reductions in landscape diversity lead to lower faunal diversity in intensively managed agro-ecosystems than in more diverse agricultural systems or in natural habitats. For example, Robinson and Sutherland found major declines in organisms associated with farmland in Britain and northwest Europe, particularly in habitat specialists. As an illustration, biodiversity declines in bird species were related to reduced food availability in the non-breeding season.

They concluded that reduced habitat diversity was of particular important in the 1950s and 1960s, while reduction in habitat quality may be more important now. Similarly, a review of the available literature on arthropod diversity found that structural biodiversity in agricultural areas is correlated with functional and species biodiversity of the above-ground insect fauna.

Tillage

Intensive tillage leads to frequent disturbances of the agricultural landscape, increases energy loss from agricultural fields, and increases problems of soil erosion and run-off from agricultural fields. All of these factors adversely affect the quality of agricultural habitats, with significant consequences for agricultural biodiversity. When looked at corn, soybean and wheat cropping systems in the Mid- Atlantic region of the United States, they found that ground-dwelling and foliage-dwelling beneficial arthropods were least abundant, and pests were most abundant, in the simplest, most intensively managed continuous corn system.

In general, ground-dwelling species were more abundant in no-till than in deep-tilled crops. This suggests that shifts towards conservation tillage and no-till will benefit agricultural biodiversity. As discussed in Section 2, such shifts have been occurring recently in many cropping systems as farmers recognize the environmental and economic benefits of doing so.

Pesticide Use

Adverse effects of pesticide use in agriculture are well-documented. Conventional insecticides generally reduce diversity through direct toxic effects. Many of the widely used classes of conventional insecticides, including organophosphates and pyrethroids, have been shown to adversely affect a broad range of non-target species, including species of economic importance. Local extinctions are common where these insecticides are frequently used. Such insecticides have been shown to eliminate important predator and parasitoid species from agricultural systems.

In Indian cotton for example, over 600 such species have disappeared altogether. These impacts on natural enemies have been shown to lead to flare-ups in secondary pest species, some of which were not previously economically important. In a few cases, insecticides directly stimulate the population growth of non-target pest species, *e.g.* pyrethroids have such an effect on some mite and aphid species. In addition, the toxic effects of insecticides can lead to food chains effects because of decreased food availability for higher trophic levels and bioaccumulation of the insecticides. For example, organochlorine use and ingestion by earthworms has led to die-offs of birds feeding on these species. Replacing broad-spectrum insecticides with more specific, softer alternatives is necessary to avoid these impacts.

Some herbicides also can be toxic to invertebrates. However, the more important effects of herbicide use with respect to biodiversity are to reduce non-crop plant (weed) populations and weed seed production in agricultural fields. Where herbicide use is intensive, adverse impacts may be seen on various vertebrate and invertebrate species that depend upon these plant species for food or shelter. Where invertebrate populations are strongly affected, consequences for higher trophic levels also may occur.

Bibliography

A. Poshadri and Aparna Kuna : *A Handbook of Food Techno's*, New India Publishing Agency, Delhi, 2013.

A.D. Dholakia: *Delicious Seafood Recipes*, Daya Publications, Delhi, 2010.

Andrew L. Winton and Kate Barber Winton : *A Handbook of Structure and Composition of Foods*, Agrobios Publication, Delhi, 2013.

Ashok K Chauhan and Ajit Varma:*Microbes : Health and Environment*, I K International, Delhi, 2007.

Aysha Aziz Faridi:*Dairy Microbiology*, Random Publications, Delhi, 2012.

Cristobal Noe Aguilar; Juliana Morales Castro; Efren Delgado; Diana Jasso Cantu and Ashok Pandey: *Food Science and Food*

Dalip Kumar and Asmi Raza: *Agriculture and Food Security*: *Contemporary Issues*, Deep and Deep Publications, Delhi, 2011.

Doreen Virtue : *Constant Craving : What Your Food Cravings Mean And How To Overcome Them* , Hay House India, Delhi, 2011.

Harmeet Singh: *Dairy Farming*, APH Publication, Delhi, 2011.

Hema Thapar: *Food Science and Health*, Pacific Publications, Delhi, 2011.

Hema Thapar: *Nutrition and Food Science*, Pacific Books International, Delhi, 2011.

M Lakshmi Narasaiah: *Economic Growth and Food Security*, Discovery Publishing House, Delhi, 2008.

Madhulika Bhatnagar.: *Encyclopaedia of Catering Technology, Food Service and Hospitality Management, Vol. I to III*, Delhi, Anmol Publication, 2008.

Manju Malhi: *Easy Indian Cookbook: The Step-By-Step Guide to Deliciously Easy Indian Food at Home*, Viva Books, Delhi, 2008.

Manoj Negi: *Career in Dairy Farming*, Abhishek Publication, Delhi, 2010.

P R Gupta:*Dairy India 2007: Serving the Dairy Industry Since 1983*, Dairy India Yearbook, 2007.

P T Rajan : *A Field Guide to Marine Food Fishes of Andaman and Nicobar Islands*, Zoological Survey of India, Delhi, 2003.

P.C. Trivedi:*Microbes : Applications and Effects*, Aavishkar Publication, Delhi, 2009.

Piyush Sharma and Prem Ram : *A Textbook of Food and Beverage Management*, Dominant Publication, Delhi, 2013.

R S Paroda: *Sustaining Our Food Security*, Konark Publication, Delhi, 2003.

R. P. Saxena.: *Hotel Management : Food And Food Services*, Delhi, Centrum Press, 2010.

S.K. Singh: *Fundamental of Hotel Management and Operations*, Delhi, Centrum Press, 2010.

S.K. Singh: *Strategic Management in Hotel and Restaurant Industry*, Delhi, Centrum Press, 2010.

Tapeshwar Singh: *Resource Conservation and Food Security : An Indian Experience (2 Vols-Set)*, Concept Publication, New Delhi, 2004.

Tharakan : *A Young Hotelier's Guide to Food and Beverage*, Tata McGraw-Hill Publication, Delhi, 2005.

Udai Veer: *Elements of Food Science*, Anmol Publications, Delhi, 2007.

Urvashi Nandal : *A Handbook of Foods and Nutritional Biochemistry*, Agrobios Publication, Delhi, 2013.

Vishwambhar Prasad Sati: *Natural Resources Conservation and Food Security*, Bishen Singh Mahendra Pal Singh, Dehra Dun, 2012.

W. Merback and A. Veha : *Crop Science and Technology for Food Security Bioenergy and Sustainability*, L. Bona, J. Pauk, ,

W.L. Slatter, T. Krishtofferson and D.A. Seiberling: *Manual of Dairy Industry : Training and Development of Personnel for the Dairy Industry*, Asiatic Publication, Delhi, 2012.

Index